Seamless Networks

Other McGraw-Hill Titles of Interest

ADAMS, WILLETTS	*The Lean Communications Provider*
ENCK, BECKMAN	*LAN to WAN Interconnection*
GRINBERG	*Computer/Telecom Integration*
HELDMAN	*The Telecommunications Information Millenium*
KUMAR	*Broadband Communications*
MINOLI	*Video Dialtone Technology*
PECAR, O'CONOR, GARBIN	*The McGraw-Hill Telecommunications Factbook*
RUSSELL	*Signaling System #7*

Seamless Networks

Interoperating Wireless and Wireline Networks

Arkady Grinberg

McGraw-Hill

New York San Francisco Washington, D.C. Auckland Bogotá
Caracas Lisbon London Madrid Mexico City Milan
Montreal New Delhi San Juan Singapore
Sydney Tokyo Toronto

Library of Congress Cataloging-in-Publication Data

Grinberg, Arkady.
Seamless networks : interoperating, wireless, and wireline networks / Arkady Grinberg.
p. cm.
Includes bibliographical references and index.
ISBN 0-07-024844-3
1. Internetworking (Telecommunication) 2. Wireless communication systems.
TK5105.5.G74 1996
004.6—dc20 96-34043
CIP

McGraw-Hill

A Division of The McGraw-Hill Companies

1 2 3 4 5 6 7 8 9 0 DOC/DOC 9 0 9 8 7 6

ISBN 0-07-024844-3

The sponsoring editor for this book was John Wyzalek, the editing supervisor was Caroline R. Levine, and the production supervisor was Pamela A. Pelton. This book was set in Century Schoolbook by Editorial Compuvision, Mexico.

Printed and bound by R. R. Donnelley & Sons Company.

This book is printed on recycled, acid-free paper containing a minimum of 10% postconsumer waste.

CBETE

Contents

Acknowledgments

Although writing a book is usually a solitary activity, writing a technical book is helped greatly by discussions with people who are experts in their respective fields or who have a general appreciation of the subject. This book contains information on varied aspects of telecommunications, and interactions with many of my colleagues contributed to the final result.

I would like to thank Khiem Le, Pete Russo, and Joe Rizzo of Bellcore for reviewing the manuscript and providing many valuable technical and editorial comments. Their thoughtful analysis was instrumental in discovering and eliminating potential points of misunderstanding.

I greatly benefited from the discussions with Gary Brush, Don Lukacs, and Ron Baruzzi of Bellcore; they helped me to see certain important aspects of my work from different points of view.

Thanks also to Renee Berkowitz, Chuck Chen, Moshe Rozenblit, and Vijay Varma of Bellcore for answering detailed questions and clarifying important issues.

Finally, I wish to thank my employer, Bellcore, for supporting this work in spirit, and for providing an environment conducive to the exchange of thoughts and ideas.

Arkady Grinberg

ABOUT THE AUTHOR

Arkady Grinberg is Director of Wireless Projects at Bellcore, where he is involved in the interconnection of PCS and wireline environments. During his career in the telecommunications industry he has also been involved in developing switching systems, product marketing, network planning, and developing standards including the SCAI standard for CTI. Mr. Grinberg is also the author of *Computer / Telecom Integration*.

Chapter

1

Introduction

This book is about networks.

It is one of the few network-oriented books, that is, books that approach telecommunications through the "eyes" of networks. This networking view is important because of the

- Growth in the importance of networks as service has become more complex
- Need to understand modern networking issues and techniques
- Need to understand how services are affected by network capabilities and network evolution
- Need to address the next stage of network evolution

The evolution of existing networks draws them into *seamless networks*, networks that hide any service-related differences from the user. In this book we will look at different aspects of networking that must be appreciated in order to understand what is involved in the making of a seamless network. We will also get a glimpse of the future that a seamless network promises.

1.1 Convergence and Integration

Convergence is the word of today. Technological developments in hardware and software allow yesterday's problems to be approached in new ways. At the same time these new abilities open new horizons and allow the creation of revolutionary solutions. Among such solutions are those that create more widespread and user friendly services for customers. Convergence of computing and telecommunications, integration of various transmission needs (for voice, data, etc.) into a single transmission medium, and merging of communications with entertainment are all examples of new solutions based on new capabilities.

Technological development means new capabilities and new problems also appear, or, rather, new capabilities raise new issues. These issues usually involve the need to integrate different technologies which were created for different reasons and also were supposed to provide different services. Opportunities presented by integration of cable TV systems with telephone networks is one such example.

This book investigates the issues raised by the need to integrate *wireless* and *wireline* networks, that is, to create seamless networks. Included are discussions about existing and planned mobile and wireline (landline) networks. Current technology and regulations support the creation of seamless networks that provide a variety of services to different users. Seamless networks make sense economically since they can attract more users and minimize new investment. They provide a competitive advantage for the companies that build and use them over those companies that provide only one type of service —wireline or wireless.

The integration of wireless and wireline networks and the subsequent emergence of seamless networks can become an important technological achievement. It can affect society through new types of services which will overcome today's limitations in communications. The following quote from Lawrence T. Babbio, Vice Chairman of Bell Atlantic, focuses on the opportunities that the integration of wireless and wireline network presents: By integrating wired network advanced intelligence, wireless network mobility and long distance capabilities, many new combinations of wired and wireless offerings are just around the corner.[1]

If mobile (wireless) and wireline networks begin to converge, what kinds of issues will have to be resolved to achieve this integration? Exactly when and how this process will take place is hard to predict, although the author believes that it is happening right now. The goal of this book is to bring up as many of the integration issues as are currently visible and discuss them with as much detail as appropriate. (It is possible that some of the issues in making networks seamless escaped the author's attention.) The book will provide a framework to address these issues. Solutions to these issues can be different, and many of them require the cooperation of the involved parties (equipment manufacturers, service providers). Creation of standards may be

[1] *America's Network*, August 1, 1995, p. 6.

necessary in some cases, but standards usually take significant time to develop. The issues of seamless networks depend on technological and standards development, and as these change, so will the importance of one or another issue.

The book does not consider in any significant detail the convergence of telecommunications networks (wireline or wireless) with cable TV systems and does not include wireless networks such as paging or dispatching. This is done to focus specifically on two major components of the integration process: wireless (really, mobile) and wireline networks. So, in the rest of the book the word *wireless* refers to mobile communications networks—both cellular and Personal Communication Services (PCS).

1.2 The Course of Investigation

The course of investigation in this book is the following. First, the total picture of the current telecommunications environment in the United States is painted, outlining the state of technology and regulations as they apply to the wireline and wireless environment. Then each type of network is discussed—its components, network services and architecture, possible evolution, and important developments, such as the concepts that support the evolution of networks. Some background material is presented in cases where solid understanding of a particular topic or concepts is instrumental in understanding specific issues. Based on the information presented for both types of networks, integration issues are identified and analyzed.

1.3 Wireline and Wireless

What do we mean when we refer to *wireline* or *wireless* networks? Which characteristics of networks make them one or the other?

There are many aspects of networks that can be studied. *Switching* and *transmission* refer to two different network functions: making appropriate connections and transferring information from point A to point B. *Call processing* and *operations* refer to the ability, mostly of software, to control (establish, clear, forward, etc.) calls and network elements for administration, maintenance, and management purposes. *User information exchange* and *signaling* refer, respectively, to transfer of user-related information (speech, data) and signaling required to control call establishment and manipulation. *Network equipment* and *customer premise equipment* (CPE) refer to the differences in ownership: CPE (telephones, private branch exchanges, or PBXs, etc.) is owned by users, while network equipment (switches, trunks) is owned by network providers.

This book uses yet another approach: distinguishing between *access* and *network*. We define a *network* as a collection of elements necessary to provide telecommunications services. These elements vary depending on the type and size of a network involved but generally include switches, transmission equipment, operations systems, adjuncts, and databases. Network elements evolved from their first applications and are considerably different now. Some of these network elements did not even exist initially and became needed with the introduction of new technologies and services.

We define *access* as an ability to connect to networks, that is, to receive services from networks. When a user wants to be served (e.g., wishes to make a call), the user indicates this desire to a network and receives an indication that this service can be provided. In the most familiar voice world this is accomplished through a user's action of taking a telephone off-hook and through a network's action of issuing a dial tone. The dial tone signals that the network is ready to provide a voice service, that is, that the access to the network has been granted. Network access is accomplished through access equipment.

Access equipment includes instruments that allow users to communicate (i.e., telephones, computers, and other CPE) and access transmission lines that connect users' instruments with networks. Access transmission lines are usually referred to as the *local loop*. If a local loop is a wire of any type (copper, fiber), the network access is wireline. If a local loop consists of radio transmission equipment, the local loop is wireless. The exception to the last statement is when a wireless local loop supports a fixed telephone. In this case the user does not even know that the radio instead of a wire is used to access the network, and the services this user receives are the same as with wireline access. Actually, this situation may be treated as the beginning phase of one type of integration or convergence. We will talk more about it in Chap. 11.

For the purposes of this book, a *wireline* network is a network with a wired local loop; a *wireless* network is a network with a wireless local loop.[2]

---

2 Although in the case of wireless access the term *local loop* is not quite correct because mobility of a wireless user defies locality, we will use this term in the book in order to use existing terminology.

Chapter

2

Current Environment

The current state of affairs in the telecommunications industry is characterized by two distinct types of networks: wireline and wireless. Their existence stems from the same basic need to communicate, but, due to different technological achievements required to support these communications needs, wireline networks have had a much longer life in the public sector than wireless networks.

2.1 Wireline Networks Evolution

The history of the wireline network begins with Alexander G. Bell and is well known. The stages of wireline networks development went from very simple arrangements for local communications to a more complicated switching environment and, then, to even more sophisticated network services.

Wireline networks serve users through a series of network elements and lines and trunks interconnecting them (see Fig. 2.1). In this kind of configuration a user is connected to a network via a hard wire over which all necessary communications between the user and the network take place (voice, data, and signaling). Since their inception, wireline networks have been concerned with the basic services—call establishment and clearing. The number of services and their complexity grew over time, requiring networks to adapt to these new services through technological innovations. Services such as call forwarding and call waiting require more sophisticated switches—the main network elements supporting users' needs.

The history of switches is basically the history of networks. First switches were simple and handled a small number of customers. With the computerization of switches and digitization of transmission and

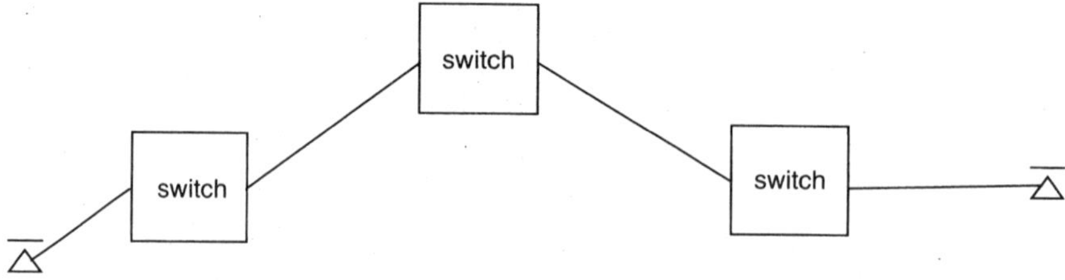

Figure 2.1 A simple wireline network of switches.

switching, switches became faster and "smarter." Computers that began controlling switches were capable of performing more complicated tasks and serving more customers. All kinds of new services that required information retention and retrieval quickly became a reality. For example, a simple call forwarding service allows a user to indicate to the network that calls to his or her telephone number should be routed to some other number instead. A sophisticated call forwarding service allows a user to indicate to the network a list of forwarding requests, something like the following:

- Calls after 5 p.m. are to be forwarded to number XXX.
- Calls from number XYZ are to be forwarded to voice mail.
- Calls between noon and 1 p.m. are to be forwarded to number YYY.

At one time, computers that control switches could be programmed to do all of the above and more. However, two types of constraints appeared. One was the cost of developing new features. Software development is expensive as is the production of huge software releases with the necessary debugging and testing. The other constraint was the length of time required to produce new software for a switch. The result is a long and costly development of each new feature, or set of features, for switches. These are just some of the reasons for the creation of the intelligent network (IN) concept.

In the IN, in addition to switches, other network elements play a pivotal role in providing services to customers. These are different computing platforms that retain specific customer-related data and other information required for a particular service and associated algorithms required to use this information. In the IN, switches do not make all the decisions. In specified instances call processing is suspended in a switch and an assigned computing platform (a database) is queried for information. When such information is received by a switch, call processing proceeds.

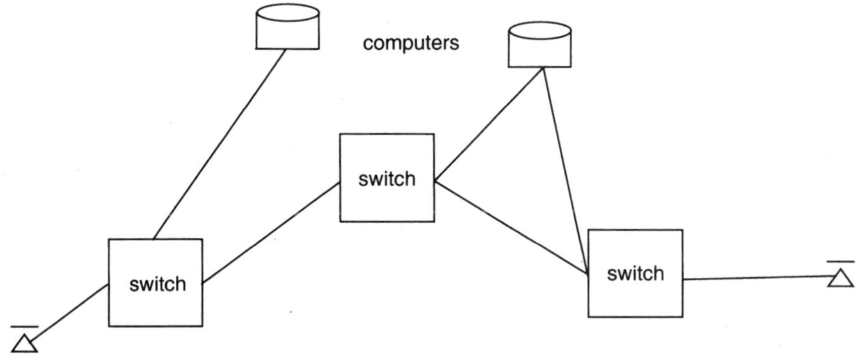

Figure 2.2 An intelligent network of switches and computing platforms.

This conceptually simple arrangement (see Fig. 2.2) allows non-switch-based computers to control some services. In our call forwarding example above, software on such a computer would understand the customer's request and issue specific directives to the switch regarding forwarding the calls. IN allows the creation of new services software not for switches but for separate computing platforms. This software is divorced from the switch development process and, therefore, does not affect switch-specific enhancements. Deployment of software in these platforms should be a lot cheaper than for switches because of faster development and testing. Another promise of IN is rapid service creation—the ability to create services on the computing platforms (databases) more or less at any time by either the owner of this database or by a third party (e.g., a software house) writing necessary software for a new service. This would speed up service introduction and make software development cheaper and a company deploying this technology more competitive.

This slight detour into the field of IN gives a background for understanding where the development of wireline networks stands, what kind of technology is being deployed, and the reasons for its deployment. Current wireline networks support a large number of users, provide basic voice and data services and services beyond basic call establishment (so-called supplementary services: call waiting, call forwarding, call conferencing, and such), and begin to provide more of the enhanced services (e.g., voice activation services). They do all this with a high degree of reliability and availability. IN technology plays an important role in today's wireline networks as does modern digital transmission and signaling systems like Signaling System #7 (SS7).

2.2 Wireless Networks Evolution

The history of wireless networks is significantly shorter. Although radio communication has been used by the military and for some civil applications for a long time, the kinds of services that we call *wireless* are actually relatively new. Since the 1960s, commercial mobile and, later, cellular networks began to appear in the United States. The first networks were very modest in terms of coverage and services.

Over time, cellular networks and the services they provided changed significantly. There were improvements in voice quality, handset sizes, and battery service time. Automatic roaming was introduced; it allows users to move from one area to another without service interruption. Additional services were introduced: data communications, voice messaging, and some of the supplementary services that the wireline networks support (e.g., call forwarding).

Currently, wireless networks are cellular networks. That is, they operate on the frequency around 800 MHz. Two cellular carriers are licensed to operate in every region using 25 MHz of spectrum each. These two operators are called A- and B-carriers. Usually a B-carrier is a local wireline network operator, and an A-carrier is a nonwireline operator. There are 734 cellular regions in the United States: 306 Metropolitan Statistical Areas (MSAs) and 428 Rural Statistical Areas (RSAs). These networks have had significant growth in their subscriber base and now serve about 28 million customers. The growth continues, and very soon new customers will have more choices in selecting their wireless carrier since Personal Communications Services (PCS) is moving in.

PCS is a new wave in wireless communications in the United States. It carries a promise of sophisticated personal services, something like UPT,[1] but it also has a distinct regulatory flavor since PCS licenses are to be auctioned to higher bidders who are to comply with the rules of deployment and service for PCS networks. A specific spectrum was allocated for PCS in the 1800-MHz band[2] in six blocks for a total of 120 MHz. Blocks A and B are 30 MHz each and are to be used within 51 Major Trading Areas (MTAs). Blocks D, E, and F are 10 MHz each and are to be used within 493 Basic Trading Areas (BTAs), and block C is 30 MHz for use within BTAs. The MTAs licenses have been auctioned off and the winners are busy planning their deployment and designing their networks and services. The auction of the BTA licenses started at the end of 1995.

---

1 Universal Personal Communications; see Chap. 14 for the definition.

2 1800-MHz and 2-GHz band are short ways of referring to the same PCS spectrum.

The latest Federal Communications Commission (FCC) spectrum allocation includes 20 MHz for unlicensed use. This spectrum is from 1910 to 1930 MHz and can be used without a license. The requirement is to conform to the etiquette rules adopted by the FCC.

In short, the current wireless environment can be characterized as (1) being based on the modern switching and transmission techniques, (2) migrating toward intelligent networks, (3) increasing in competition for market segments in both wireline and wireless areas driven by new technologies and new regulations, and (4) being an unstable state of affairs in the wireless industry due to allocation of new spectrum and a period of preparation for new PCS-driven competition that will definitely change the landscape of wireless networks in the future.

Chapter

3

The Wireline World

The wireline world consists of different types of networks, all having the same characteristic: the users are connected to these networks via hard wires. The wires can be different (copper, fiber), but all of them provide a hard medium over which electrical signals travel between users' equipment and networks (see Fig. 3.1). User equipment can be a telephone, a fax machine, a key-system, or other such devices. Having a hard wire is useful because, in practical terms, there is not much interference with the electromagnetic waves traveling over the wires and conveying the necessary information. However, wires also have disadvantages, namely, high costs of labor during installation and maintenance, and inflexibility (the inability of equipment at the user's end of a wire to move in space). It is where it is, and services that this equipment can provide are tied to its rigidity.[1]

This chapter addresses different network types and their components. It gives background information for the concepts that are important in today's wireline networks and gives examples of implementations of these concepts. As a short summary of the state of affairs in wireline networks, the following can be said: Wireline networks provide great call connectivity on demand using sophisticated protocols, such as Signaling System #7 (SS7). A variety of supplementary services is also provided, the most well-known being call waiting, call forwarding, and three-way calling. Intelligent networks (INs) are being introduced, and with them more services are being offered. INs are also providing the foundation on which sophistication of services

[1] A partial exception is a cordless phone, which allows its users to move about, but not too far. There is also a question whether cordless phones, the way they are used today, belong in the wireless or in the wireline world.

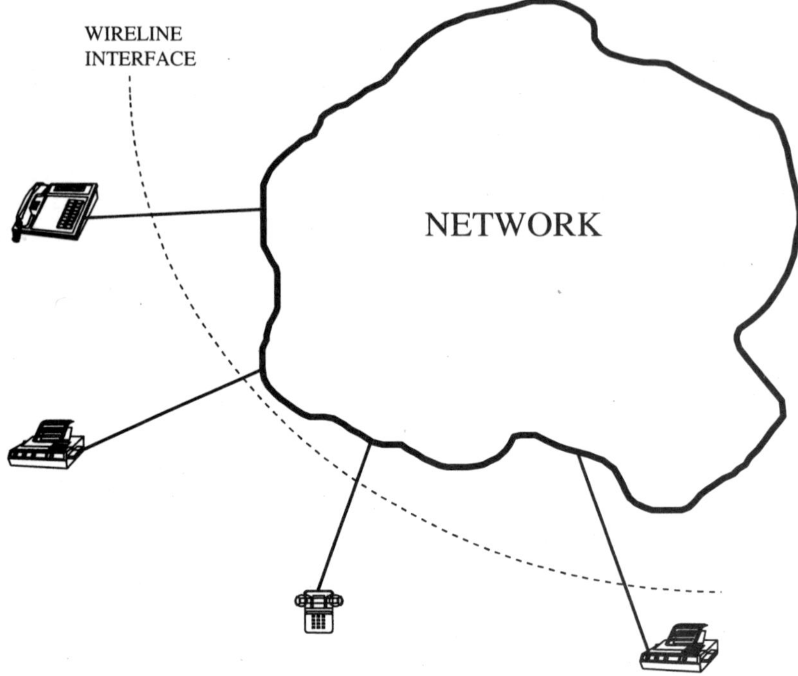

Figure 3.1 Wireline connectivity.

will grow exponentially. The modern wireline networks are gearing up to fully use their digital capabilities in conjunction with new services supported by INs.

3.1 Network Types

Networks can be grouped into public and private; public networks, in turn, can be grouped into local exchange carriers (LECs—local networks) and interexchange carriers (IXCs—long distance networks). In general, public networks serve all the public, and private networks serve specific groups of people (e.g., employees of a large corporation).

Local public networks are those that serve local customers only. Locality is defined by existing regulations. After the divestiture of AT&T, the MFJ[2] dictated the establishment of Local Access and Transport Areas (LATAs) throughout the United States. Each LATA (or, as it is sometimes referred to, service area) is identified by its geo-

[2] Modified Final Judgment (MFJ)—the document governing the AT&T divestiture.

NJ LATAs: 220 - Atlantic Coastal
222 - Delaware Valley
224 - North Jersey

Figure 3.2 LATA example.

graphic boundary. A local telephone company (LEC) serves customers within each LATA that is under its control. Figure 3.2 shows three LATAs which cover New Jersey. Each LATA has a number and a name. In the case of the three New Jersey LATAs, the numbers are 220, 222, and 224. Very small portions of Pennsylvania are also contained within one of these LATAs, but usually LATAs occur within state borders. Within a LATA calls are established (or connections are established) over a relatively short distance.

Most U.S. users are connected to local carriers. Examples of LECs are Nevada Bell and United of New Jersey. If a call is originated in an area and is destined to another area that is not part of the original local area, this call has to be "passed" to an IXC, which will handle this call between two LECs. Even if for a given call the originating and terminating localities are served by the same local network but these localities belong to different LATAs, an IXC must handle the connection between them. In Fig. 3.3 three types of calls are shown. Call 1 is a local call and is handled by an LEC. An example is a call within the same city. Call 2 originates and terminates within the boundaries of the same LEC but involves two different LATAs; therefore, an IXC handles the call between LATAs A and B. For example, if you call from Flemington, New Jersey, to Atlantic City, New Jersey, an IXC is involved in completing the call. The distance is not that large, but the call crosses the LATA boundary and must be handled by an IXC. Call 3 originates and terminates in different LATAs within different LECs and involves an IXC. For example, a call originates in New York City (NYNEX), terminates in Atlanta (BellSouth), and is handled by an IXC (a long distance company, e.g., AT&T) between these two states.

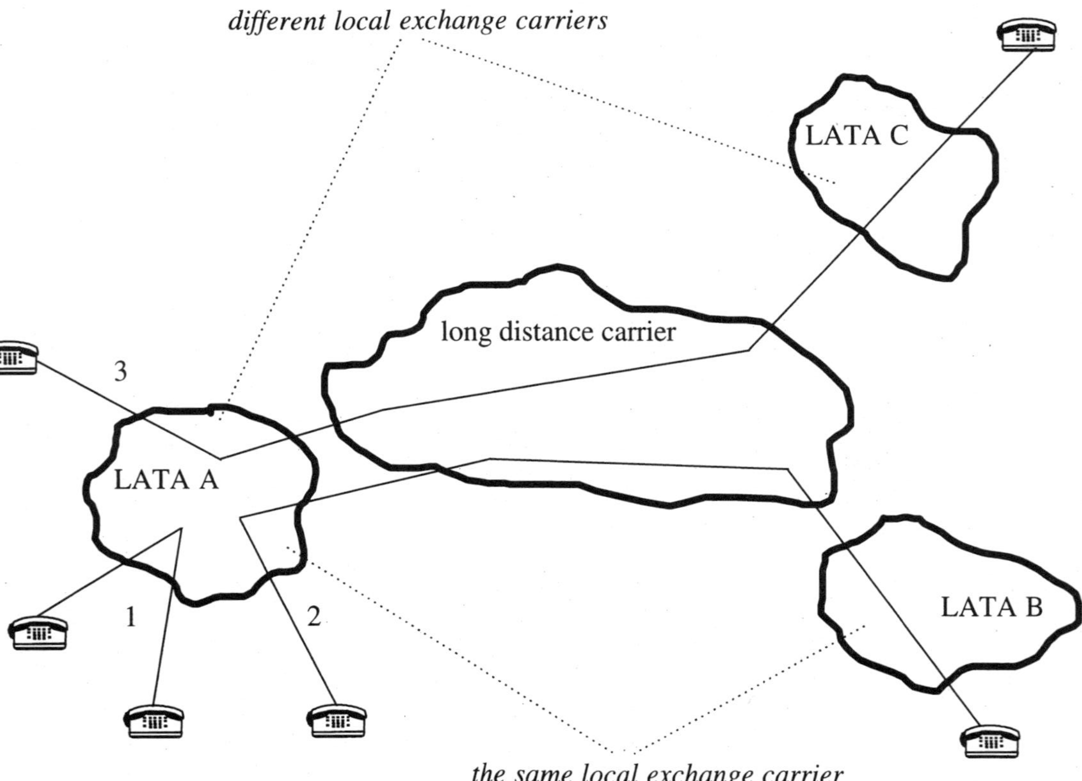

Figure 3.3 Call handling by different carriers.

Private networks, as we said before, serve a specific group of customers, such as employees of a company or a state government. A typical private network is either a localized network, for example, serving students and employees of a college, or a more global network serving, for example, employees of a company located in several cities and, even, countries. In short, an institution builds a private network for its private use, and the network's facilities are not for use by outsiders (see Fig. 3.4).

Both public and private wireline networks are built with the same principles in mind, and although there are differences, they use similar technology. For example, switching systems perform all call processing functions,[3] and switches used in public and private networks are similar but not the same. The private branch exhanges (PBXs) used in private networks are usually smaller than their public networks counterparts (although some PBXs are very large), and they usually support a somewhat different set of services required by a particular group of people using a specific private network.

Although there are service-related, regulatory, and other differences between various network types, in the rest of this book we will refer to all of them as wireline networks. Since the subject of this book is the integration of wireline and wireless networks in general, distinguishing networks by their access is sufficient for our discussion. Therefore, networks with the same kind of access, over wires, are considered together.

3.2 Pre-IN Networks

Wireline networks have been serving people for many years using basically the same technological approach: Switches are completely in charge of establishing, clearing, and otherwise manipulating calls generated by users. Transmission equipment and operations systems are also used (see Fig. 3.5).

The role of switches is a central one. They recognize call requests from users, assign necessary resources, route calls to appropriate destinations, provide information required for billing, etc. Local loop and trunk transmission facilities contain different equipment required to operate access lines and interswitch trunks. This equipment consists of various termination electronics, cross-connects, multiplexers, and other such items. Operations systems (OSs) are computers executing specific complex tasks, such as provisioning, maintenance of various

---

[3] At least in the pre-IN networks this is true. But even with IN, switches are the basic network elements which perform actual call *switching*.

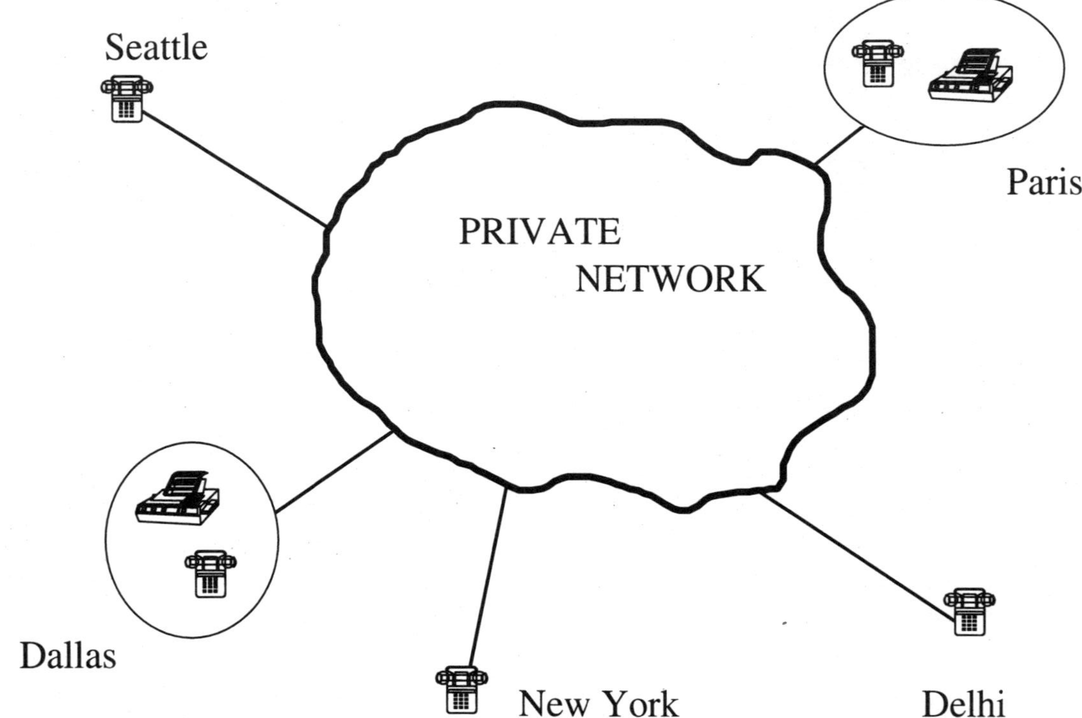

Figure 3.4 Private network.

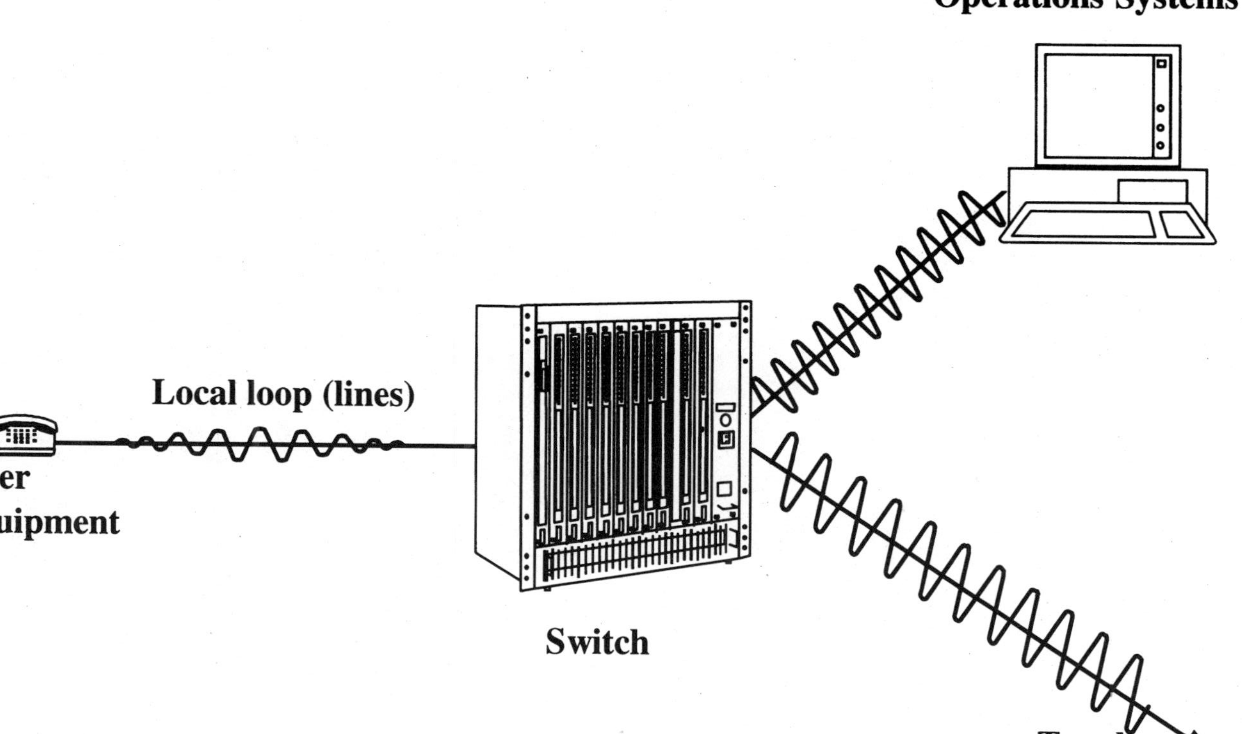

Figure 3.5 Basic pre-IN network.

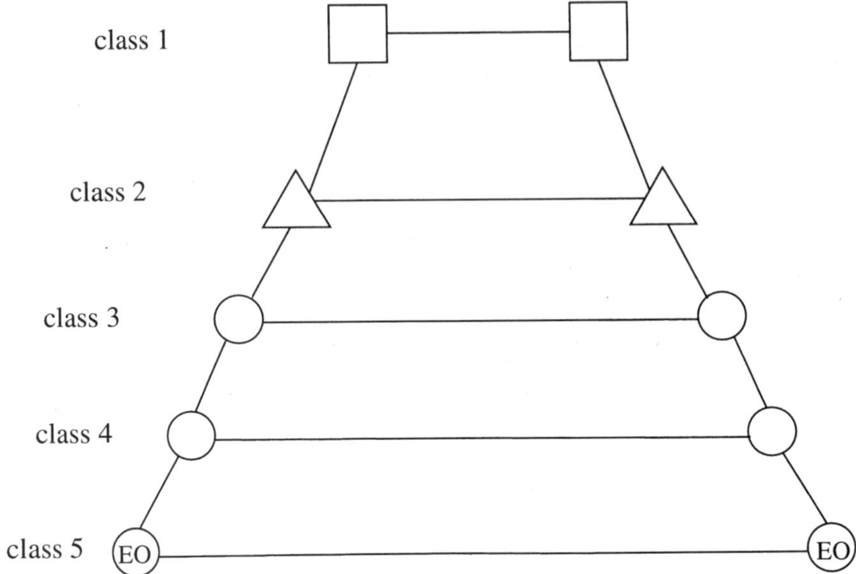

Figure 3.6 Switching hierarchy.

network elements, and traffic collection. Although OSs are not the subject of this book, they are a critical element of any network operation and, therefore, an important part of any network architecture.

To complete an architectural picture of the pre-IN wireline network, it is useful to mention the network hierarchy that has existed since the days of the AT&T monopoly. The hierarchy is built to optimize traffic patterns and, as a result, use equipment more efficiently and economically. It consists of several levels of switches, each level responsible for call connectivity over a specific area: local area, state, region, or country. Figure 3.6 shows the network hierarchy. A routing plan is defined for each network; according to this plan routing can be accomplished by accessing nonadjacent levels of the hierarchy. The routing plan is designed with the optimization of network resources in mind.

The lowest level in the hierarchy is an end office (EO),[4] or class 5 office. For each originating call an end office is responsible for finding a route to a destination EO. The hierarchy supports a variety of routes depending on specific traffic circumstances. If for any reason no direct route between an originating and a terminating EO is available, the

[4] Another widely used term is *central office* (CO).

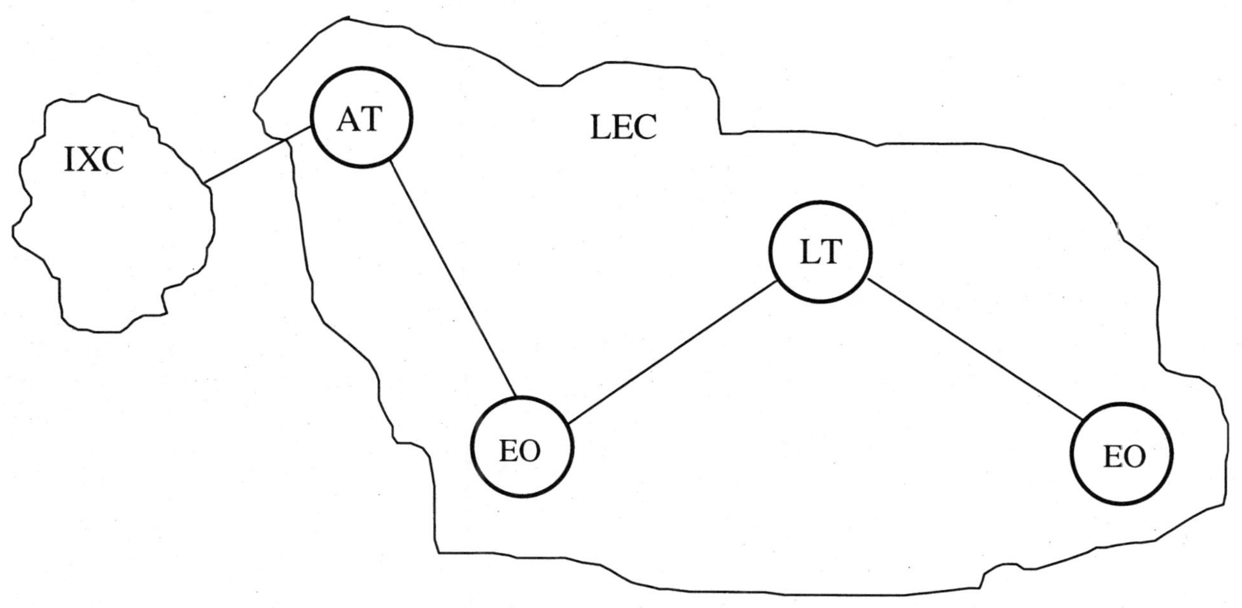

Figure 3.7 Tandems.

call is passed up to the next level, which is a class 4, or a toll center (TC). The same considerations apply at this level of the hierarchy. Any call can reach a class 1 office if this route is the most appropriate.

Only the lowest level of switches (class 5) is involved with direct interconnection to the user equipment. Other levels are only involved in call routing and connection establishment.[5]

Switches of classes 1 through 4 usually belong to IXCs. Class 5 offices belong to LECs. In addition, LECs also own tandem switches that help to streamline traffic. Tandem switches perform two types of functions: as an access tandem (AT) and as a local tandem (LT). The AT functions as a traffic concentrator between an LEC and an IXC. The LT concentrates and distributes traffic within a local area (see Fig. 3.7).

As mentioned, this hierarchy was created when a single company controlled the network end to end. The hierarchy was used to optimize network performance and was easily applied to specific changes in traffic and service conditions. Since a single network does not exist any more, each network operator builds its own architecture. The terms of the hierarchy we presented here are familiar in the industry and are very useful in discussing network architectures. However, the number of hierarchy levels used by each IXC may be different. Some may use all four levels, although more likely only three or even two levels are used, depending on specific conditions. It is safe to say that the LECs rely on a two-level hierarchy, with tandem switches playing the role of the higher level. It will be interesting to look at the networks and their hierarchies when regulatory restrictions on providing local and long distance service change or disappear. Will that mean a return to the original five-level hierarchy?

3.2.1 Services

Wireline networks support many different services. These services vary from a local public network to a long distance public network to a private network. However, the idea of this chapter is not to list every possible service supported by every network or even by a majority of networks. Instead, it will discuss the kinds of capabilities that the majority of wireline networks offer as services to their customers.

It is not simple to address network services because what constitutes a service depends on who is buying. For example, public net-

---

[5] This statement was absolutely true before the current telecommunication environment was created. Today, long distance carriers can connect their class 4 switches (second level) to the user equipment. However, for long distance carriers, class 4 switches constitute the lowest level, and, therefore, for the purposes of this book, we can still say that only the lowest level of switches is involved with direct user equipment connections.

works lease lines to users, and these lines constitute a service to those users. But these kinds of services do not exemplify the types of network capabilities that make today's wireline networks what they are. Therefore, we will focus on the services that characterize these networks.

The most obvious of these services is call establishment on demand. Today's wireline networks use very sophisticated switching and transmission equipment, along with modern signaling and computers. This equipment supports connectivity throughout the civilized world (by civilized, I mean the countries where telephones are reasonably widespread; this is not a social definition, but a telephony one). The availability of a dial tone, which is a prerequisite to making a call, is almost immediate and voice quality is excellent most of the time. These are crucial characteristics of wireline networks. All new services, including those provided by wireless networks, are compared to them. People in the United States are accustomed to hearing a dial tone as soon as the phone is picked up. Voice quality is also being taken for granted.

Services which are provided in addition to the basic call establishment are usually referred to as supplementary services or enhanced services. Many of these services are made possible by the availability of SS7, whose messages carry information regarding calling and called parties that can be used to support many features. Other services are made possible by the enhancements in the switches themselves. Among the supplementary services are the following:

- *Call forwarding.* The ability to specify where (to which particular destination) to forward calls when certain criteria are met, such as when the destination is busy or all the time.
- *Call waiting.* The ability to indicate to a user who is on a call already that another call is incoming. The user can then switch between the two calls by pressing a flash button. This service requires the switch to be capable of connecting two different calls to the same destination.
- *Cancel call waiting.* The ability to cancel call waiting at any time, so the destination will appear as busy to a newly arriving call.
- *Call screening.* The ability to monitor calls that have been forwarded to another destination.
- *Three-way calling.* The ability to originate two calls from the same telephone and connect all three parties together for a conference call.
- *Call hold/retrieve.* The ability to put a particular call on hold and retrieve this call after a certain time.

Some services are being introduced under the common name CLASS (Customized Local Area Switching Services). These are more sophisticated services, although some of them are enhancements to the existing supplementary services. They are:

- *Call waiting deluxe.* The ability to act in different ways on a call which is waiting: keep the call on hold, forward the call to another designation, play an announcement, etc.
- *Calling number delivery.* The ability to receive the number of the calling party (before deciding whether to receive it or not)
- *Calling number delivery blocking.* The ability to block the delivery of a calling number to users who subscribe to calling number delivery service
- *Automatic callback.* The ability to call the last number dialed without redialing
- *Distinctive ringing.* The ability to indicate incoming calls with different ringing patterns depending on the calling party and other circumstances

These are just some of the supplementary services that wireline networks provide today without the support of intelligent network capabilities. Other types of services include the following:

- *Voice messaging.* A very popular type of service that allows callers to leave messages in the special voice mailboxes to be later retrieved by the owners of the mailboxes. Many different options exist for voice messaging, depending on the needs.
- *Data communications services.* These provide the ability to exchange data between different computing devices. Data services are either packet- or circuit-switched and allow different transmission rates, such as 1.2, 2.4, 9.6, and higher up to 64 kbps. Even higher rates are available if appropriate loop technology, such as fiber, is used.
- *Centrex.* This is a generic name for services that a public network provides to its customers that make the customer's environment appear to be a PBX. Centrex is a functional PBX. It offers basic and many additional services similar to those offered by a PBX (although a PBX typically offers a greater variety of additional services). Subscribing to Centrex alleviates the need for a capital to buy a PBX. Centrex services are used by the same type of organizations that purchase PBXs.

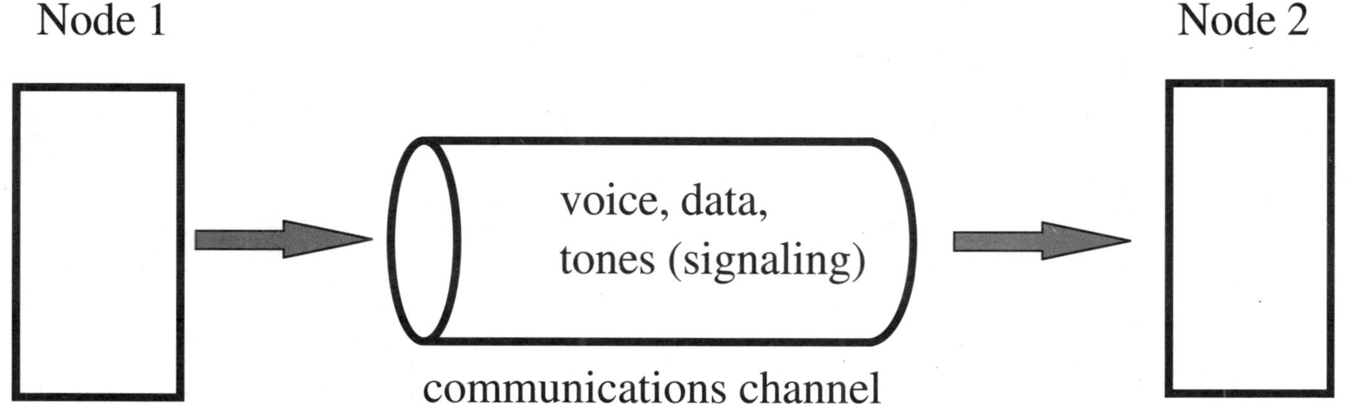

Figure 3.8 In-band signaling.

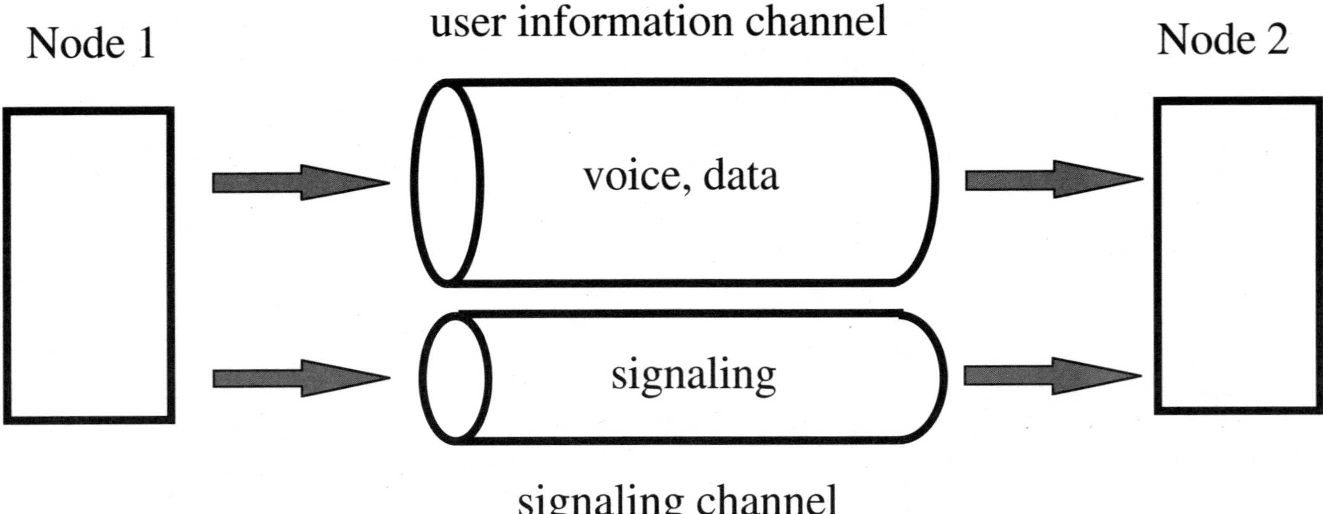

Figure 3.9 Out-of-band signaling.

- *Emergency service (911).* This is a public safety feature that allows a free emergency call from anywhere in the country; in most instances, the identification of the caller is also made.
- *Operator services.* These provide support for coin phones, directory assistance, alternate billing, and other similar services.

3.2.2 Architecture

The architecture of the pre-IN network is generically shown in Fig. 3.5. In such a network a switch performs all call processing functions aided by transmission facilities and controlled by operations systems.

In this architecture all signaling is *in-band*. In-band signaling is the technique of exchanging all signaling information over the same communications channel (band) which is also used for the transmission of user's information (e.g., voice or data). In in-band signaling systems electrical impulses of a specific type and duration are usually used to indicate some control information. For example, in such a system dialed digits are passed as dual-tone multifrequency (DTMF) tones within the channel that will later be used for voice communications (see Fig. 3.8 on p. 23).

Another type of signaling is called *out-of-band*. This signaling technique implies the use of a channel specifically designated for the exchange of signaling information (see Fig. 3.9). The user's communication channel is freed to be used only for the user's needs (voice, data). The signaling channel is used to transport signaling information in a digital form. Protocols in the form of messages are specifically designed for the purposes of a signaling channel. These messages convey data between involved systems. An example of such an out-of-band signaling protocol for network access is Digital Subscriber System 1 (DSS1), the Integrated Services Digital Network (ISDN) access signaling protocol.

Common channel signaling. Within the network out-of-band signaling is usually referred to as *common channel signaling* (CCS). A specific protocol—SS7—has been developed and standardized by international and domestic standards committees. SS7 is used all over the world to support network signaling needs. No modern network can function without CCS, and SS7 is the preferred protocol. Let's first look at the network architecture with SS7 that is shown in Fig. 3.10.

This figure differs from Fig. 3.5 by the addition of network elements required for SS7. These elements are:

- Service switching point (SSP)
- Signal transfer point (STP)

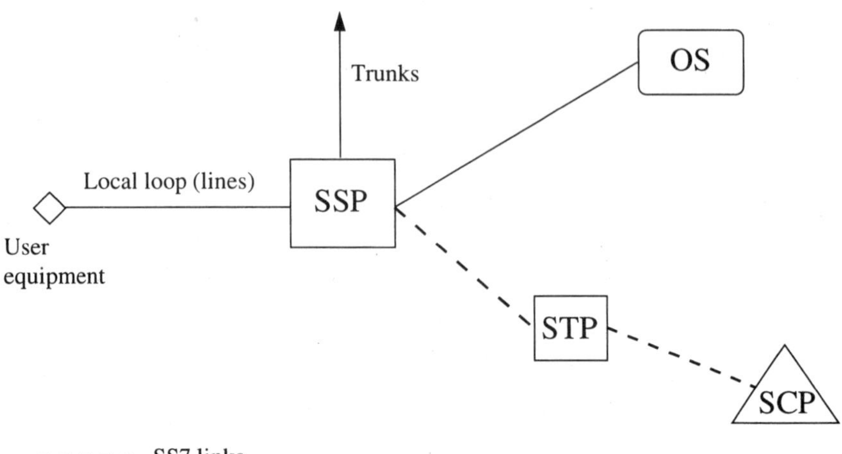

Figure 3.10 Pre-IN network architecture with CCS.

- Service control point (SCP)

An SSP is a switch equipped with SS7 capabilities, an SCP is a database performing number translations (e.g., for the 800 service) or other such functions, and an STP is a packet-switching system which routes SS7 messages (packets of signaling information) among various SSPs and SCPs. A typical network configuration using SS7 is shown in Fig. 3.11.

Two aspects of the SS7 network are worth mentioning here: redundancy and different link types. Redundancy is an important aspect of SS7 networks. SS7 has been designed with robustness in mind. Since SS7 serves the entire network and carries information crucial to network services, its reliability is of paramount importance. Failures of SS7 elements can bring a network down if it is not designed properly. To increase reliability and availability of SS7 elements, they are usually designed to be redundant. Switches (SSPs) are very expensive network elements and designed to carry certain traffic and support certain services. Redundant switches would be very expensive; therefore they are never redundant.[6] STPs, on the other hand, are usually redundant (see Fig. 3.12). They are deployed in mated pairs; that is, instead of a single STP, a pair of them is installed. Both elements of a mated pair share the load, but each one is capable of handling the full

[6] Switches and other network elements usually are built with internal redundancies. That is, depending on the product, certain key elements are designed to be redundant so the overall system is more reliable.

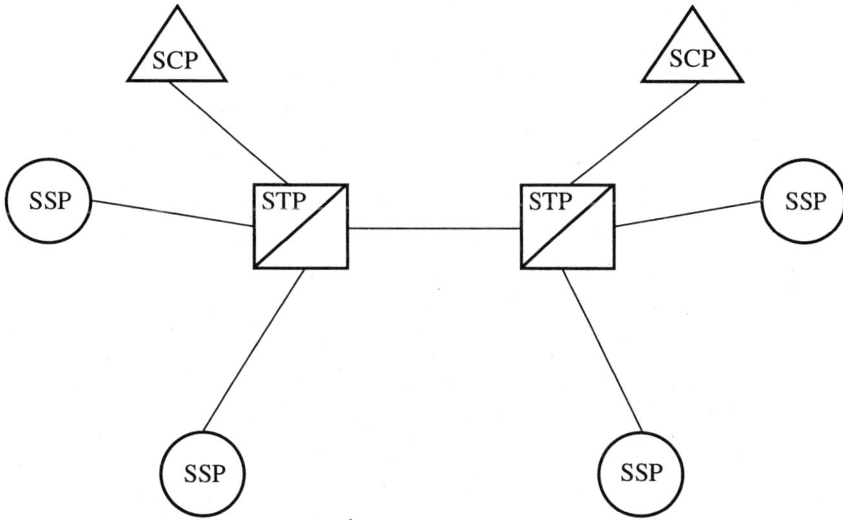

Figure 3.11 SS7 network.

load expected for this network area. Two elements of a mated pair do not have to be in the same location. An additional degree of survivability is achieved by locating them remotely. For example, in earthquake-prone areas placing one STP remotely from its mated STP may allow a network to survive an earthquake if it occurs in one location but not in another. Since STPs are used by many SSPs, failure of any of them can bring a large portion of a network down. Certain links are also usually made redundant, for example, A links (discussed below). In addition, link redundancy is supported by the deployment of redundant STPs and SCPs.

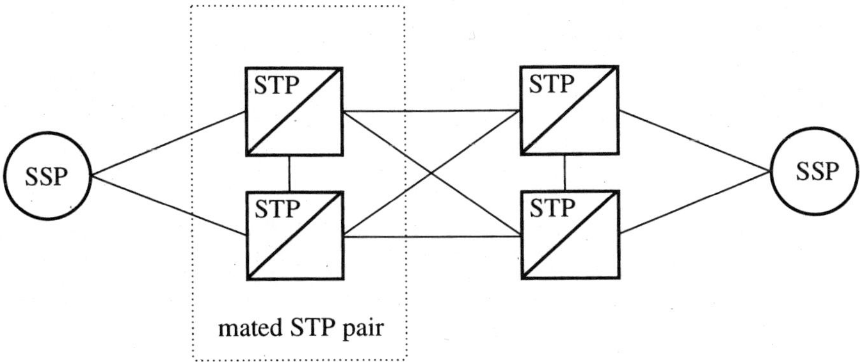

Figure 3.12 SS7 redundancy.

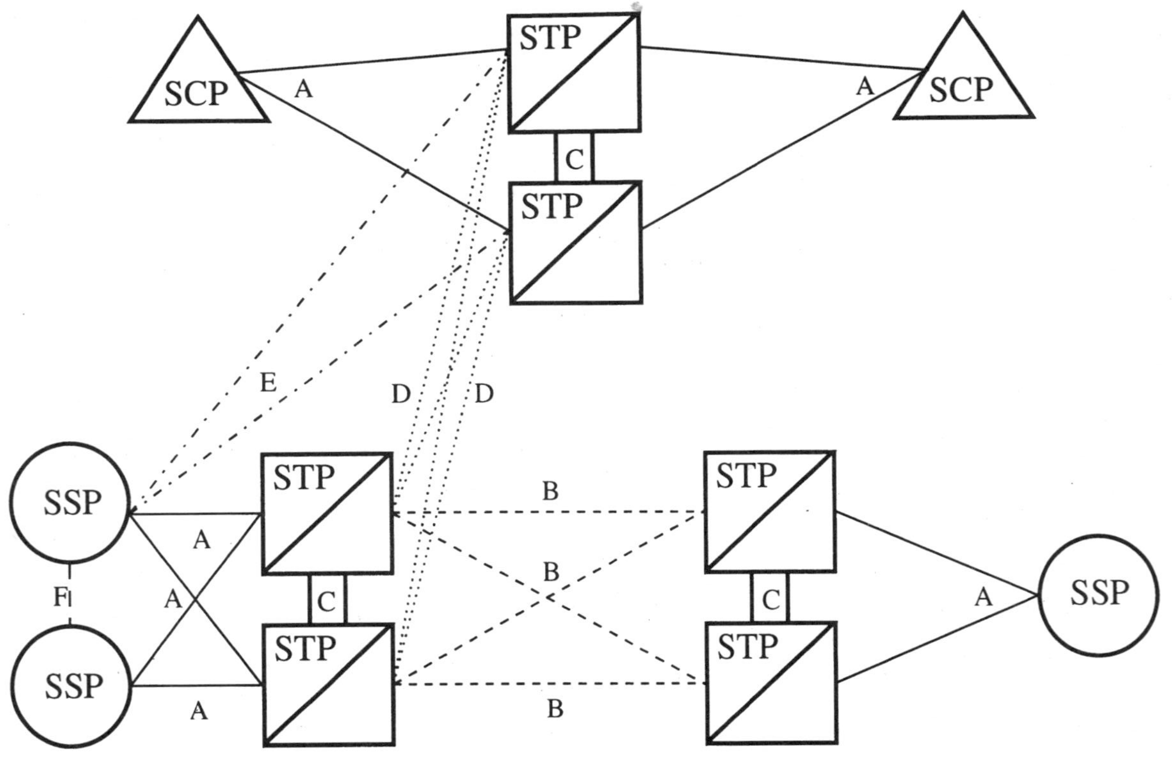

Figure 3.13 SS7 links.

SCPs are also usually redundant because they retain critical information. However, the SCP operations and the links between mated SCPs are not defined in the SS7 standards. An installation of a mated SCP pair requires careful considerations of the synchronization procedures between mated SCPs. Various deployment schemes can be used for mated SCPs, such as a load-sharing configuration, and hot or cold standby configurations. This issue is decided by individual network implementors.

Now, let's discuss signaling links. Different links are used between different pairs of network nodes. The significance of link names is in the usage of specific links. When SS7 networks are designed or performance models are built, link names carry information about their use. The following are the defined SS7 links (see Fig. 3.13):

- *A links (access links).* Used between STPs and SSPs or STPs and SCPs. The maximum number of A links to any STP is 16, and to any mated STP pair it is 32.
- *B links (bridge links).* Used to connect mated pairs of STPs at the same hierarchical level.
- *C links (cross links).* Used to interconnect two mated STPs. These links are used for signaling traffic only in the case of congestion and unavailability of other routes.
- *D links (diagonal links).* Used to interconnect mated STP pairs at different hierarchical levels. They are used only when different STPs are deployed at different levels.
- *E links (extended links).* Used between SSPs and remote STPs for signaling route diversity.
- *F links (fully associated links).* Used between SSPs when signaling traffic volume justifies associated signaling between SSPs. This requires some explanation. In SS7 associated signaling is defined as the signaling mode in which the messages conveying information between adjacent nodes are sent over a link directly interconnecting those nodes. In a quasi-associated signaling mode the messages are sent over links that traverse at least one more node which is not either the origination or the termination node (see Fig.3.14).

Each node in the SS7 network has a point code—an address by which it is distinguished from other nodes. To communicate (to send messages from node to node) the point code of a destination has to be known. The possible number of nodes in a network is bound by the number of bits in the point code field of the protocol, which is quite large in the United States—24 bits.

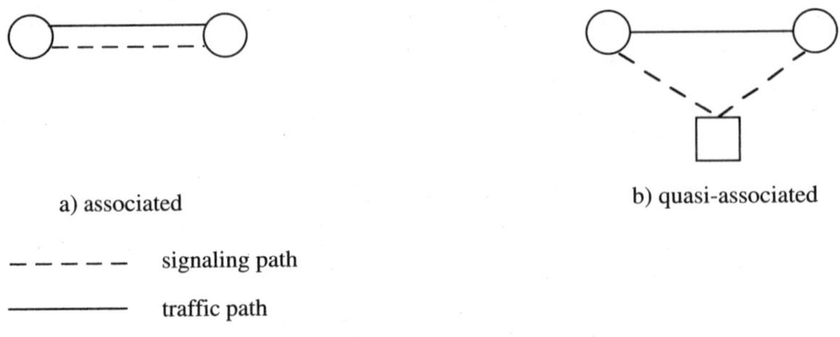

Figure 3.14 SS7 signaling modes.

Now we need to discuss the protocol architecture of SS7 since it is important to understand in relation to IN applications. SS7 is actually a set of protocols, called parts, which work together to provide specific functions. Not all of the SS7 parts need to be used at the same time. Their use depends on the specific needs of the network. Figure 3.15 shows the protocol architecture of SS7, which consists of the following parts:

- Message transfer part (MTP)
- Signaling connection control part (SCCP)
- ISDN services user part (ISUP)
- Transaction capabilities application part (TCAP)

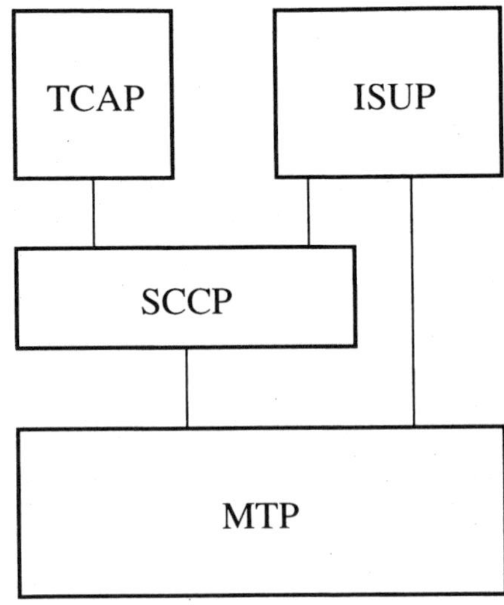

Figure 3.15 SS7 protocol architecture.

MTP is responsible for the transfer of SS7 messages between network nodes. It provides routing, error detection and correction, and message sequencing. MTP is only concerned with the transfer of information over a link to an adjacent node.

SCCP is responsible for addressing and routing in general. It provides a function, *global title translation*, which is used to find a real address of a node based on some system-related name—a global title. It is global because it is recognized within the entire network. A global title is used for flexibility in routing. Global titles allow knowledge of point codes to be centrally located. Otherwise, every node in a CCS network would have to know the address of every other node. Global titles are translated into real addresses of the nodes by STPs. Translation is a process of mapping global titles to true destination addresses. Global titles are translated based on their types. A type identifies a translation table to be used. For example, dialed digits can be used for this purpose and translated into a destination address.

ISUP is responsible for call control. It sends and receives messages between network nodes that convey information for call establishment, call clearing, etc. When an ISUP message is constructed and ready to be sent to a destination, it uses either SCCP/MTP or just MTP for message routing and delivery.

TCAP is responsible for non-call-associated information exchange. While ISUP's functions are call-related, TCAP's functions are defined for other needs, such as transfer of non-call-related information. In general, TCAP is a protocol created for transaction-oriented activity. It represents a specialized application of a more general protocol defined by ITU-T, called Remote Operations Service Element (ROSE).

Other application parts exist within the SS7 suite of protocols, but they are not important for our discussion. Examples of these are Telephone User Part (TUP), a pre-ISDN call control protocol, and Operation, Maintenance, and Administration Part (OMAP), which deals with operations, maintenance, and administration.

Now we need to spend some time understanding TCAP because it plays a key role in the functioning of the Advanced Intelligent Network (AIN) protocol.

TCAP.[7] Since TCAP is an extension of ROSE, we will start the discussion with ROSE functions and capabilities. ROSE is a collection of functions that support invocation of operations remotely; that is, it

7 Because of TCAP's relation to the OSI concepts and the application layer structure in particular, it would have been better to discuss TCAP after those concepts are presented. But TCAP is an SS7 protocol and should be part of the SS7 discussion. Please see the appropriate text about Open System Interconnection (OSI) later in this chapter.

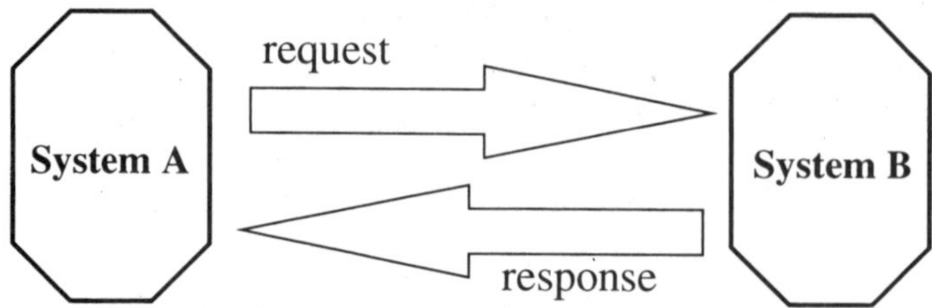

Figure 3.16 The ROSE model of remote operations.

enables remotely located systems to exchange information (commands, data) that results in the invocation of specific actions. In short it can be said that ROSE supports distributed applications. The model that explains ROSE is very simple (see Fig. 3.16). It implies that a request issued by a system to another system may generate a response from the other system back to the originating system. Five classes of services are defined (see Table 3.1). Every time a remote operation is initiated, it is of one of the five classes, depending on the circumstances.

ROSE defines the following services it provides to the application that uses it:

- INVOKE
- RESULT
- ERROR
- REJECT-U
- REJECT-P

INVOKE is a service that an application uses to request an action to be performed by a remote system. When a request is to be issued,

Table 3.1 ROSE Operation Classes

Operation class no.	Definition
1	Synchronous, reporting success or failure (result or error)
2	Asynchronous, reporting success or failure
3	Asynchronous, reporting failure only
4	Asynchronous, reporting success only
5	Asynchronous, outcome not reported

the ROSE INVOKE service is used. The response to a request can be of one of the four types indicated above. When the response indicates a successful outcome of the request, the RESULT service is used. Otherwise, either an error is reported, or the request is rejected either by a service user (REJECT-U) or a service provider (REJECT-P). A service user is an application that uses ROSE, and a service provider is a provider of the ROSE service.

The ROSE standard identifies basic services that can be used by any application that requires support for remote operations. TCAP is built according to the same model for remote operations, but while ROSE provides a general truck for carriage of any information in support of any operations, TCAP specifies which operations it supports. These operations are specifically defined for the SS7 environment.

TCAP messages consist of a transaction portion, a component portion, and the newest addition—a dialog portion. We will not discuss all the elements of each portion because the intent here is not a TCAP tutorial but rather its brief description in order to understand the important functions of TCAP.

The transaction portion of a TCAP message identifies the package type, which can be one of the following:

- Unidirectional
- Query With Permission
- Query Without Permission
- Response
- Conversation With Permission
- Conversation Without Permission
- Abort

A particular package type is used depending on the requirements of a given transaction. It identifies the nature of the transaction.

A Unidirectional package is used when information is just being sent to the destination system and no response is needed.

A Query-type package is used to access remote information. A query implies a reply, and it can be provided in different package types. Two types of Query packages are defined: Query With Permission and Query Without Permission. Permission refers to the ability of the application receiving a query to end the transaction. If permission is granted, the receiving application can end the transaction.

A Conversation-type package is used to maintain a dialog between the two systems which has been started by a query. Two types of

Conversation packages exist: Conversation With Permission and Conversation without Permission. The indication of permission grants the receiver of the message an ability to end the conversation.

A Response package is used to end the conversation or to respond to a query. This would be the last message in a transaction.

An Abort package is used when the originator of the transaction has to end the transaction due to unexpected events.

Maintaining transaction IDs is the second important function of the transaction portion. A transaction ID is a value used as a correlation between messages that belong to the same transaction. If two systems are involved in a conversation, two transaction IDs are used—originating and responding. The initiating application chooses an originating transaction ID, and the responding application chooses its own responding transaction ID. This mechanism allows each application to maintain its own identifier for a particular transaction. When a transaction is initiated, an originating transaction ID is assigned. It is maintained during the entire transaction, while messages of different package types can be used. A responding application assigns a responding transaction ID, which is also used in the conversation until the response is issued. A sequence presented in Fig. 3.17 shows an example of such a use of a transaction ID. The transaction IDs are released after a response-type message or an abort-type message has been received.

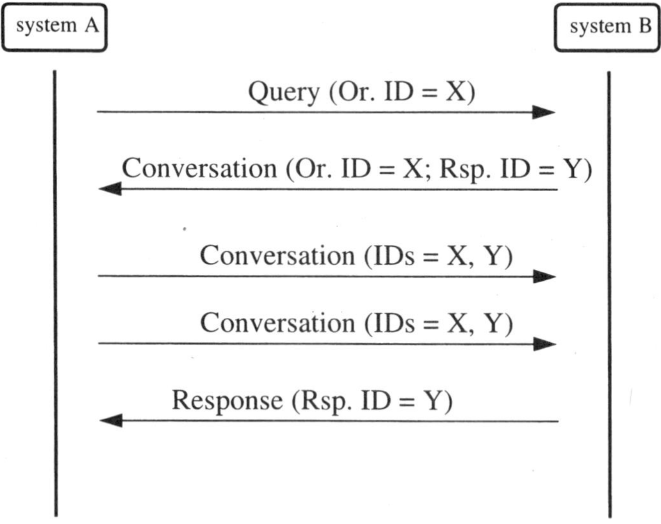

Or. ID = Originating transaction ID; Rsp. ID = Responding transaction ID.

Figure 3.17 Maintaining transaction IDs.

The component portion of a TCAP message identifies specific operations to be performed. More than one component can be packaged within the same transaction portion. Each component contains a list of parameters that explain the requested operation to be performed. The component portion closely follows ROSE and provides six component types based on ROSE's services:

- Invoke (Last)
- Return Result (Last)
- Return Error
- Reject
- Invoke (Not Last)
- Return Result (Not Last)

In comparison with the ROSE services a new qualifier appears—Last/Not Last. This indicates whether the component (Invoke or Return Result) is last in this particular transaction or not. Other than this, the meaning of these component types is the same as in ROSE.

The dialog portion of a TCAP message is the latest addition to TCAP both in domestic and international standards. Its purpose is to create associations over which a dialog between two systems can take place. The dialog portion provides two major services: exchange of application-context information and security. An application context specifies

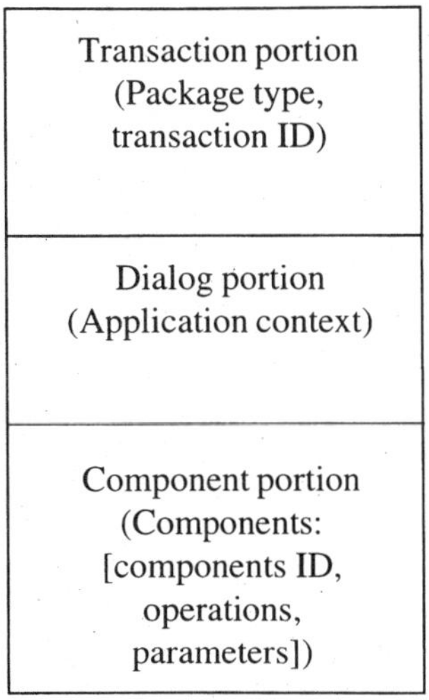

Transaction portion (Package type, transaction ID)
Dialog portion (Application context)
Component portion (Components: [components ID, operations, parameters])

Figure 3.18 TCAP message structure.

what kind of environment an application can operate in (e.g., specific functions used or a set of parameters recognized by this particular application context). Security services provide functionality to maintain a secure application-level association between peer applications.[8]

The functions of the dialog portion are modeled based on another existing ITU-T standard—Association Control Service Element (ACSE). As in the case of ROSE, TCAP followed the concept of ACSE but has the same functions used within TCAP.

The TCAP message contains many elements. Figure 3.18 on p. 35 shows the general structure of the message and identifies only the key elements that we discussed.

Access interfaces. Wireline network access interfaces, although all are wires of some sort, are different. The most widely used are 2- or 4-wire interfaces supporting the plain old telephone service (POTS). Physically speaking, it is a copper wire with a limited bandwidth. This kind of interface served telephone networks since their inception but is slowly being replaced by other media capable of transmitting more information in digital formats. One such interface is DSS1 to support ISDN; another is Broadband ISDN (BISDN), which uses fiber optics.

Digital access interfaces allow planning for more sophisticated services because they are capable of supporting more complex signaling procedures. ISDN and, especially, BISDN provide users with a mechanism to access more services in the network.

Another access type is through cable TV. This access technique uses coaxial cable to pass information between a user and a network. While a cable has more bandwidth than a copper wire, its use as an access to a full-service telecommunications network is still under consideration. Currently, most of the cable plant serves cable TV customers; that is, a cable TV provider transmits signals only one way. Incorporating cable into telecommunications services is actually another aspect of integrating different environments, in this case telecommunications and cable TV environments.

3.2.3 Numbering

Telephone numbers are the means to address specific destinations—people, computers, and others. Numbers are a finite resource. Hence, numbering is important for business since without numbers that can be assigned to subscribers, one cannot have subscribers.

---

[8] See the discussion of Application Service Elements (ASEs) later in this chapter.

In the United States numbers are assigned by North American Numbering Plan Administrator (NANPA)[9] using a specific format: N0/1X-NNX-XXXX, where

N is any digit 2 through 9.

0/1 is either 0 or 1.

X is any digit.

The first three digits in the format are usually referred to as Numbering Plan Area (NPA). It identifies a geographic area in the country where the number belongs. NNX identifies a specific central office. XXXX is one of the extensions in the NNX central office. So, a typical number in the United States would be something like 214-678-1234.

With this scheme there are limits to the available numbers; this is already beginning to affect certain areas of the country. The total number of NPAs using the above format is only 152 (160 total minus 8 reserved numbers of N11 format). To help the situation, the NPA format has been changed into Interchangeable NPA (INPA), which is NXX. Since the second digit can now be any number, the total number of NPAs is 792 (800 total minus 8 reserved).

Another option is to increase the number of central offices per an NPA. The original format of central office codes was NNX. By changing this format to NXX, the total number of central office codes is increased from 640 to 800.

3.3 Wireline Networks with IN

During the 1980s, a new concept was developed in the evolution of networks: intelligent networks. Why this name was chosen is not clear; it is not that networks without IN are lacking the intelligence required to complete calls or to provide other services. IN basically redistributes functions and makes things more suited for the introduction of new services but does not really make networks more "intelligent." So, what is IN?

3.3.1 IN drivers

At some point in the development of switching systems, they became very large and costly in terms of software required to control all functions. Every new software release, usually produced to handle new

---

9 Bellcore has been performing the functions of NANPA. However, it has announced its intention to relinquish this responsibility, and it is possible that when this book is published, NANP administration will be performed by some other company.

features, would cost a lot of money and take a long time to develop. Operators of networks, the people who buy and use these switches, wanted to find a simpler and faster way of introducing new software for new services. The speed of service introduction may make a big difference in the competitive position of a network operator. This idea translated into a concept of distributed intelligence in the network, that is, instead of switch-centralized intelligence, it would be distributed according to some rules. Some of the decision making would be retained in a switch, while other decisions would be made not in a switch but in supporting network elements. In addition, this concept promised a shortened software development cycle in comparison with the switch software development. It also promised an ability to introduce services faster due to a shorter development cycle and the use of an innovative service creation environment that allows a rapid development of new features based on already available functional components.

At the same time technological evolution reached a point that would allow the development of this concept into a practical application. Computers required to run IN applications were becoming faster and cheaper, capable of addressing more memory and storing more information. Networks were migrating to digital transmission, simplifying the task of developing a signaling system to handle interactions between various computer-based network elements. Networks were also being equipped with a modern signaling system on which the IN concept would be built. The road for IN was open.

3.3.2 Services

What makes IN services special?

Are there services that can be supported only with the IN platform?

These are interesting questions that require some thought to answer. No answer can easily satisfy everyone working on services. However, I hope the view expressed below may be a consensus one.

IN services are supported in the special computers (SCPs) that contain controlling logic and necessary data. Services are triggered in the switches by occurrences of predefined events, and these computers are consulted for actions that are to follow. Generally speaking, switches could also be equipped with service-related logic, and they would provide the same services the SCPs in IN provide. Differences do exist, though. First, as we discussed in the beginning of the IN sections, a distributed environment provides an opportunity to modify and add services faster and more cheaply, thereby providing an incentive to introduce new services. Second, often IN services can be supported with more efficient use of resources. Centralized decision-making and database look-up make switches concentrate more on their main tasks—call switching and control. So, the main characteristics of IN services in comparison with non-IN services are their efficiency and relative independence from switches.

The answer to the second question depends on how far forward into the future we need to look to satisfy the inquiry. When the IN platform was developed, the definition of IN services started with existing services, such as CLASS (available at that time). Obviously, the IN designers wanted to make sure that the new platform could handle services already supported by the networks. After that new services were added, but most of them, from the functional point of view, could be accomplished in the switches without IN (albeit less efficiently, in most cases). These are the services that are currently being supported by the AIN specifications. Future service development will be based on the already available IN platform and the expertise that people will have developed working with it. These new services may be of the type that could not be supported without IN, at least not efficiently. The kinds of services that would fit into this category are those that require region or networkwide intelligence. Positioning such services within switches would not be effective.

The logical conclusion of this discussion should be that the IN services are the same as those before IN with the understanding that new services will appear, especially when Universal Personal Communications (UPT) services begin to emerge. However, there are already a number of services being defined as IN services and, more importantly, these are the services associated with IN. One can say that it is irrelevant whether the same services can be provided without IN or not: They are treated as IN services. Among these services are the following:

- *Outgoing call restriction.* The ability to disallow certain outgoing calls; different criteria for restricting calls can be used.
- *Ring control.* A feature used by the voice messaging service subscribers to control the number of rings before the call is passed to the messaging service provider.
- *Voice-activated dialing.* An ability to initiate calls by voicing a request instead of using a dialing pad.
- *TelePages.* An enhanced directory assistance.
- *Teledemocracy.* An interactive voting feature.
- *Call center services.* A number of different services associated with various incoming and outgoing services that require interactions between switches and computers, such as telemarketing, attendant desk, and others.

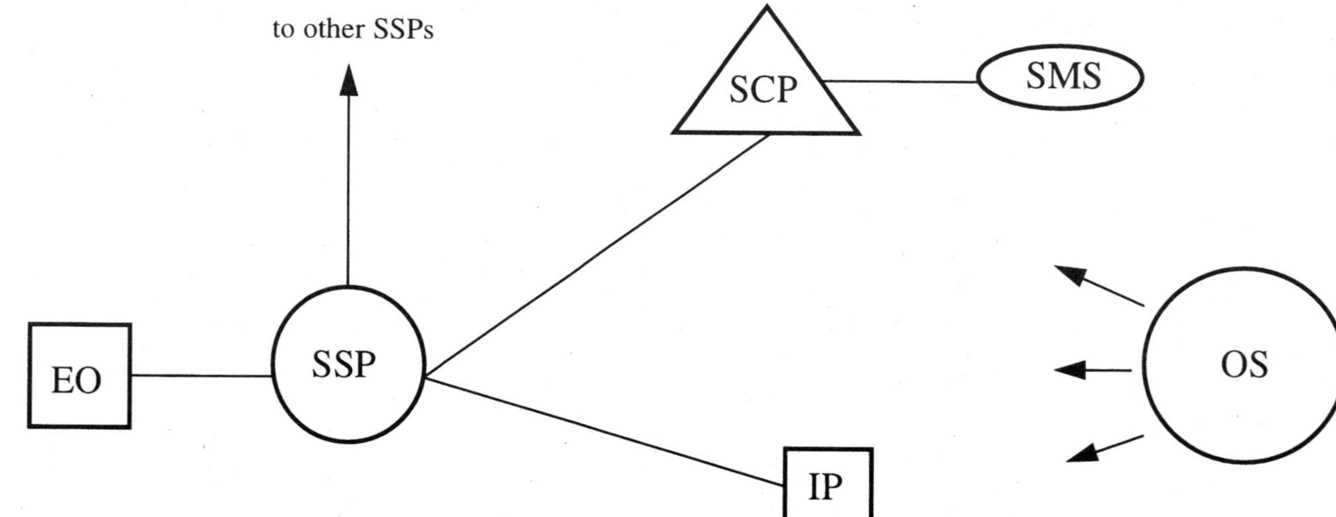

Figure 3.19 IN architecture.

3.3.3 Architecture

IN is based on two technological developments: CCS and layered protocols, with the highest layer directly supporting applications. The architecture of IN is derived from the architecture of the signaling network, and the main protocol for carrying information within IN is SS7. Although the architectures of SS7 and IN look remarkably similar, it would be a mistake to think that they are one and the same. They were created for different reasons. SS7 was developed as a digital CCS system—a carrier of information between network nodes to support various network signaling needs. IN was developed as a platform for distributed intelligence (information and control). IN uses SS7 as a mechanism for information transport, but its benefits materialize with applications, which are software programs that perform complex tasks in support of services offered to users.

Figure 3.19 shows the IN architecture. The elements of the IN architecture are as follows:

- Service switching point (SSP)
- End office (EO)
- Service control point (SCP)
- Intelligent peripheral (IP)
- Operations system (OS)
- Service management system (SMS)

SSP performs switching functions. We already used the term *SSP* before when we discussed the SS7 architecture. This is basically the same element, but with the addition of IN functions. These functions recognize specific predefined events and act on them by suspending the call processing and issuing requests for instructions or information. The act of performing one of these functions is called *triggering*, and parameters that are set to invoke triggering are called *triggers* (see more on triggering later).

The EO is also a switch, but with no IN functionality. For call treatment using IN capabilities, a call originated at an EO is passed to an SSP.

The SCP performs functions that can be characterized as database-related. An SCP is a computer which contains applications and associated databases. Again, this is the SCP that we met while discussing the SS7 architecture. The IN SCP is basically the same but is open to many more applications as it has evolved from a service-specific node to a service-independent node. It is also the place where the service creation would be performed. How service creation is to be accomplished is not specified in standards in as much detail as other aspects of IN, but the SCP is the place to do it.

OSs perform all necessary network administration, operation, and maintenance functions. They are connected to all network elements. The SMS is specifically provided to work with an SCP. It is the SCP's provisioning system.

Information between the SSP and SCP is transported using SS7 TCAP. The interface between SSPs is SS7 ISUP.[10] The interface between IPs and SSP is not well defined because of the relative lack of common definition for IPs. These systems are defined to do whatever their creators and service providers want them to do. They can be designed to perform voice activation functions, announcements, etc. The IP was intentionally allowed to be ambiguous to fit the needs of a particular service provider. However, because of this ambiguity it is difficult to define a single interface capable of supporting all known and unknown needs. Bellcore defined an ISDN based interface between the SSP and IP based on some known needs of public networks. Different interfaces can be supported for different needs.

Figure 3.19 shows the IN architecture as it was initially developed. While the architecture remains stable, another element has begun to appear—a service node (SN). A service node is not a well defined network element. Its functions depend on a particular company's views. Some envision an SN as a combination of a number of different IPs, others see it as a combination of switching and database functions, and still others see it as a combination of all of the above. Many considerations determine how people define this new IN element: costs, optimization of performance, etc. At the time of this writing, no stable view of SN functionality has emerged, and we cannot discuss it on equal level with other IN elements. However, this element will play an important role in the discussions in Chapter 9.

Triggering. The IN architecture distributes intelligence among different network components. The result is that a switch needs to communicate with an SCP to complete many activities. The ability of a switch to do that is provided by triggering. Triggering is a mechanism by which call processing is suspended in a switch while it querries an SCP for certain information. Switch call processing resumes once the information is received from the SCP (or after a specified time-out). The mechanism consists of trigger detection points (TDPs) which are

---

[10] Use of ISUP or TCAP also amplies the use of underlying protocols, either SCCP and MTP or just MTP. While TCAP can be used with other lower-layer protocols, like X.25, ISUP is only used with SS7 lower-layers.

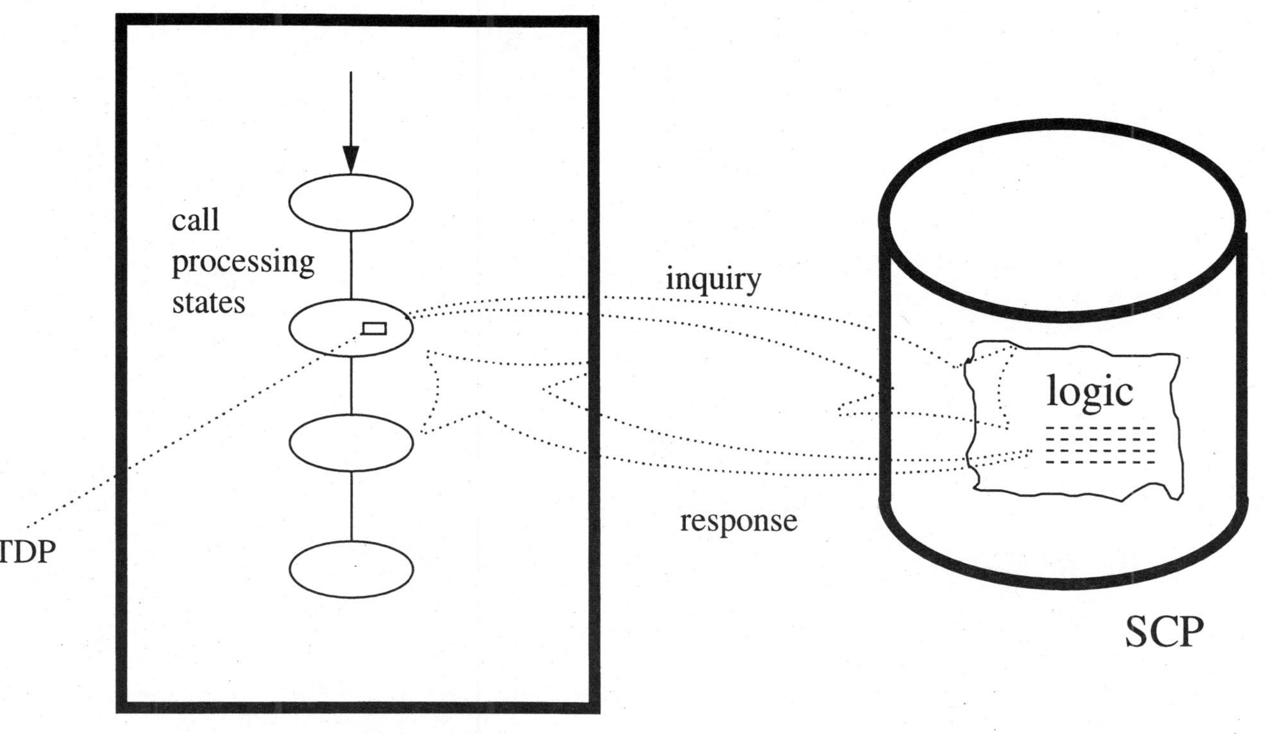

Figure 3.20 Triggering in IN.

inserted in specific places in call processing, messages generated when a trigger is activated, and the logic located in an SCP that responds to the messages activated by triggers. In general the mechanism works as described in the following example (see Fig. 3.20).

A call is coming into a switch (SSP). It is handled by the SSP according to the call model, represented here as a series of call processing states. In a state where digit analysis is performed, a TDP is located. A trigger associated with this TDP is set to be activated by NPA = 800 (just as an example). If the incoming call contains a called number starting with 800, the trigger is encountered. The result of a trigger being activated is an inquiry (a message) from an SSP to an SCP with a full 800 number. A special logic (application) in the SCP translates this number into a real directory number (DN) and sends a response (another message) with a DN to an SSP, which completes call routing using the DN.

This is a simple example, but the actions of triggering are the same for other cases. The information received from an SCP may control routing, specify services, or direct an SCP to perform specific actions. A number of TDPs have been defined in AIN and, by knowing which particular TDPs are implemented in SSPs, one can deduce which services can be supported by particular equipment.

Service creation. Rapid service creation is one of the main drivers of IN. The goal is to establish an environment in which modification of existing services and creation of new ones would be simple, relatively speaking. The existence of a system which is a service hub with an open access to its software would support the idea of rapid service creation. The role of such a system is assigned to an element of an IN which was mentioned several times—an SCP. An SCP is the place where all control information resides (control information = intelligence). An open interface to an SCP with the goal of modifying or creating services would, therefore, complete requirements for IN service creation. This is accomplished through a service creation environment (SCE).

Currently, there is no standard for an SCE. All manufacturers involved with this element use their own judgment in designing it. A generic SCE is usually connected to an SCP either directly or through an SMS, but logically an SCE can always be viewed as being connected to an SCP because it affects the SCP's operations. Figure 3.21 shows such a logical connectivity. The arrow "manipulations" should be understood as indicating that the SCE, through its internal activity, virtually "manipulates" the processes in the SCP, achieving the creation of new processes.

Many different processes exist in the SCP to support various services. These services can be built out of elemental building blocks.

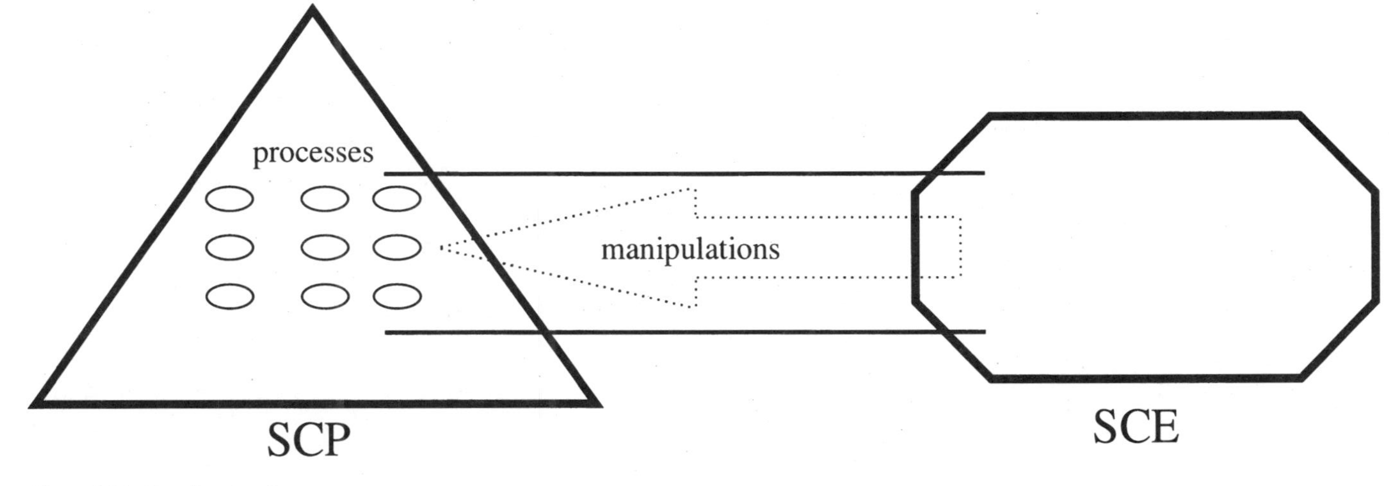

Figure 3.21 Service creation.

Suppose service *Follow* requires five different actions to take place. These actions, or elemental building blocks, are sequenced as part of the software development process for service Follow. The same building blocks may be used in a different fashion to provide another service—Follow1. An SCE would be the system capable of supporting such a rearrangement of functions already programmed in the SCP to create a new service.

Another example of using an SCE is the creation of a completely new service using the existing capability elements (building blocks) in the SCP. Let's assume that among the capabilities in the SCP are elements A, B, C, and D. Capability A is used in service 1, capabilities B and D are used in service 2, and capability C is used in service 3. Other capabilities are also used in services 1, 2, and 3, but they are not of interest to us because in constructing service 4 only elements A, B, C, and D are used. Through the SCE these four capabilities are strung together to create this new service 4.

The idea of building blocks permeates the entire IN concept, and service creation is also based on the same idea. It is not hard to imagine that many elemental capabilities are initially programmed in the SCP and later, through the use of an SCE, these capabilities are reused and rearranged to create new services. Of course, if a specific capability necessary for a new service is lacking, a new software release is required. Therefore, it is important to implement as wide a variety of capabilities as practically possible when an SCP is being developed.

After a new service algorithm has been created, the job of an SCE is usually not over. A generic SCE is capable of validation of the newly created algorithm. It would make sure that the capabilities used in the service are not strung together in an incorrect way. If a mistake or an inconsistency is found, it is flagged. This process is similar to debugging software. After the validation process is complete, an SCE performs testing. An SCE accepts certain parameters that are used in the service and tests the new algorithm. The results of testing are indicated in a form (graphical, text) that depends on the particular system.

When a new service is created through the use of IN techniques, its provisioning and activation take place as for any other service. All necessary operations systems that are involved in the service provisioning and activation process are used for a newly created service.

3.3.4 Applications and protocols

We already mentioned that IN is partially based on the concept of layered protocols with the highest layer responsible for applications support. This is a very important concept in the understanding of IN and the differences between wireless and wireline worlds.

This is the concept of open systems, that is, the systems that use interfaces defined in a manner that allows intersystem communications with any other open system. The foundation of the open system concept has been standardized by the International Standards Organization (ISO). The standards define the OSI reference model, which specifies different layers in the protocol stack and the interfaces between these layers. The ISO OSI reference model consists of the following seven layers (see Fig. 3.22):

- Application layer (7)
- Presentation layer (6)
- Session layer (5)
- Transport layer (4)
- Network layer (3)
- Data link layer (2)
- Physical layer (1)

Each layer performs specific functions and is functionally independent of other layers. When two open systems communicate, equal layers exchange information as peers. So, information created by, for

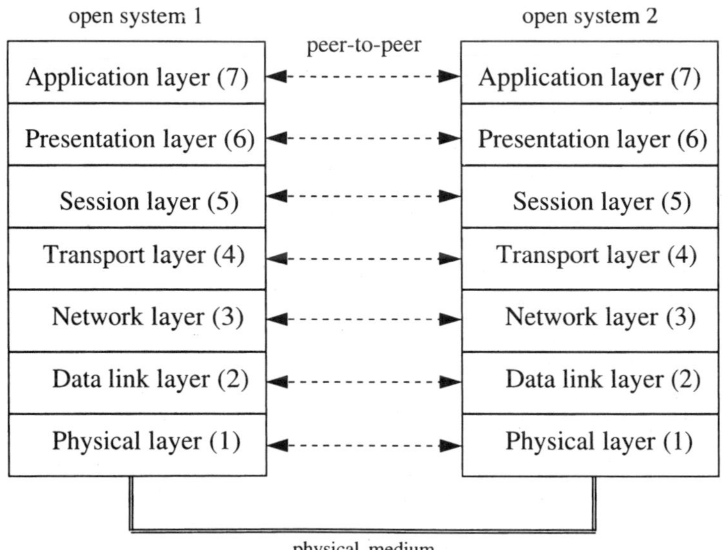

Figure 3.22 OSI reference model.

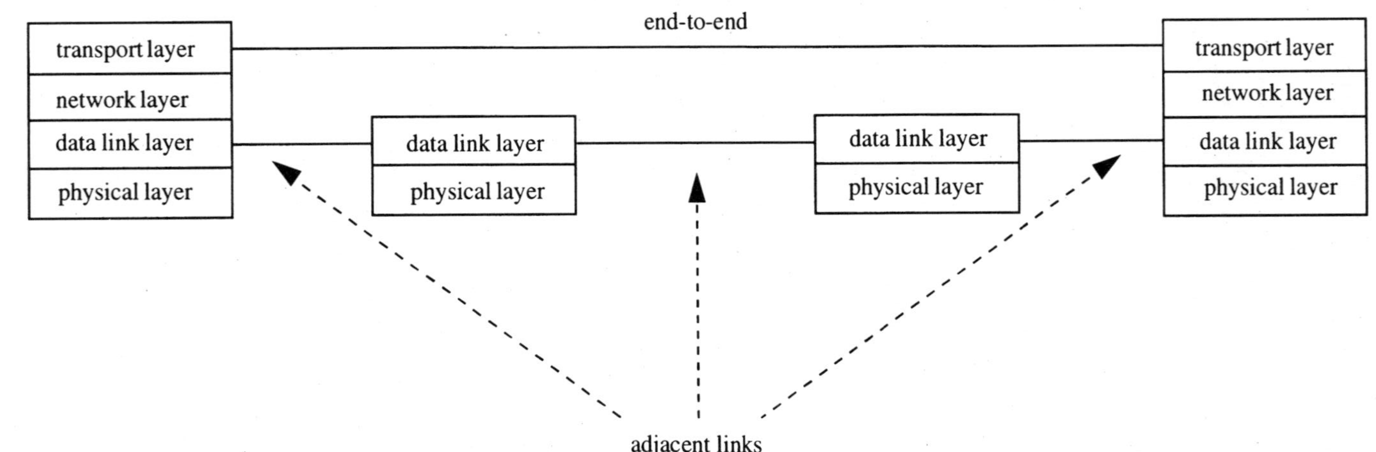

Figure 3.23 Different scope of coverage by layers 2 and 4.

example, a session layer in system 1 would be understood by a session layer in system 2, and so on.

Physical layer. Layer 1 is responsible for the actual transmission of data bits over a physical medium. Note that a physical medium connecting two open systems is not considered to be in the physical layer. Rather the physical layer uses it to transmit data.

Data link layer. Layer 2 is responsible for an accurate transmission of data between two adjacent systems. It performs the functions such as error detection, error correction, and data sequencing.

Network layer. Layer 3 is responsible for addressing and routing. It is the role of this layer to find a path from an originating node to a destination node which the data messages would follow.

Transport layer. Layer 4 is responsible for reliable end-to-end data transmission. It performs the necessary functions to assure an accurate delivery of data through the path identified by the network layer. It is somewhat similar to layer 2 in its role as a controller of reliable data transfer. However, layer 2 is only concerned with a link to the adjacent node while layer 4 is concerned with data delivery to the final destination (see Fig. 3.23).

Layers 1 through 4 are concerned with actual transmission of information between systems. They take information provided by the higher layers and assure its delivery to the destination. Layers 5 through 7 are concerned with applications themselves. They perform different functions in support of successful communications desired by an application.

Session layer. Layer 5 is responsible for the synchronization of the dialogue between two systems. It keeps track of the dialogue and can restart this dialogue if a failure in communication occurs at a lower layer.

Presentation layer. Layer 6 is responsible for the representation of information (data). It negotiates with other open systems to determine how to represent data so it can be understood. The simplest example of this function is the negotiation of the language to be used in the communications (e.g., "let's speak Esperanto").

Application layer. Layer 7 is the highest layer in the reference model and the only one directly "connected" to applications. The entire stack of layers is concerned with the communications aspects of an application; it is used only when an application needs to communicate with the peer application in another system. Layer 7, though, is unique among the layers because it is the layer that interprets the application's requests and generates communications messages. On the receiving end, the application layer receives information from a peer system and presents it to its local application.

Application layer structure. The application layer provides an interface between user's applications and the communication protocol stack. Because applications are many, varied, and fast-changing, the application layer was in itself structured to support different needs of applications. Although the structure of the application layer is quite complex and a detailed study of it is not required for the purposes of this book, it is necessary to understand some basic principles of the application layer to appreciate the benefits of IN-based and other applications. Figure 3.24 shows a simplified structure of the application layer, which should be sufficient for our further discussion.

An application is a user program designed to perform a specific task. When such a program requires communications with other software programs, it uses the communications capabilities represented by the protocol stack we are discussing.

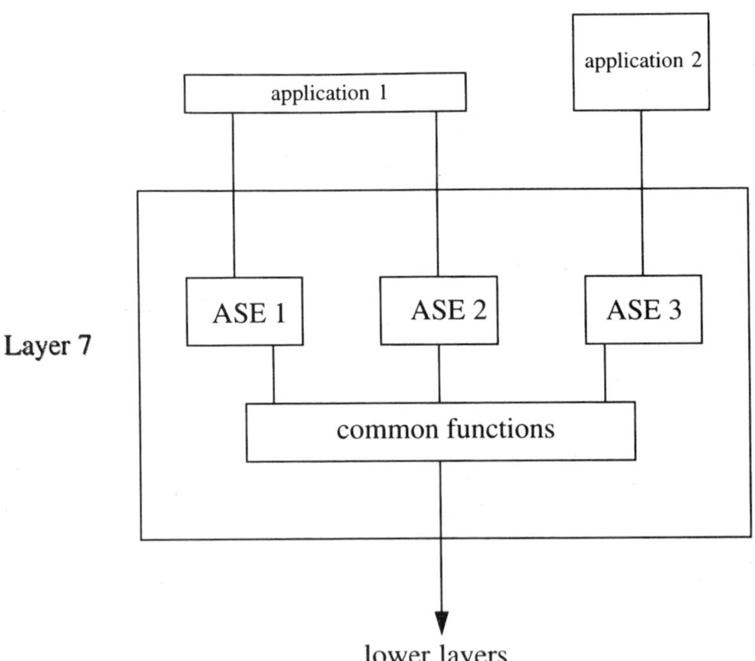

Figure 3.24 Simplified application layer structure.

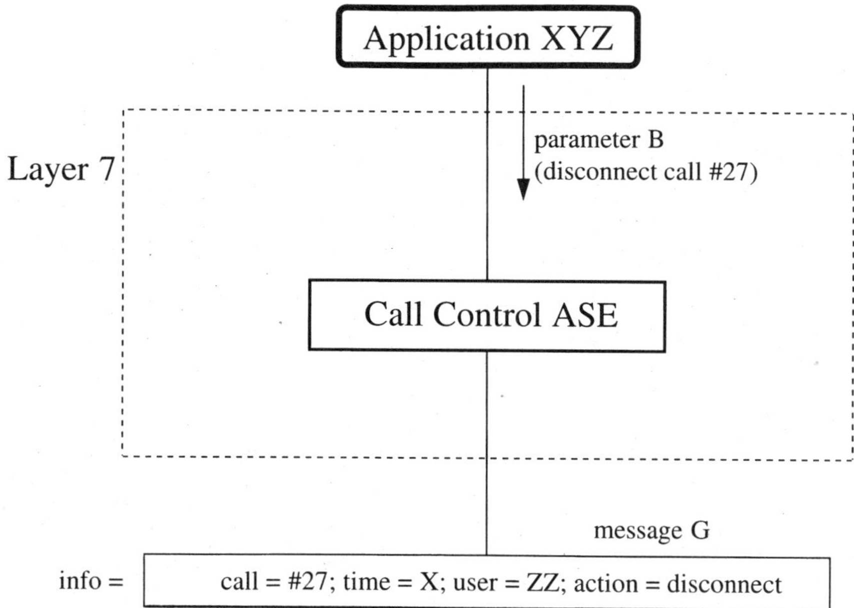

Figure 3.25 Application uses an ASE to initiate a call disconnect.

An ASE is a set of functions grouped together to support specific applications needs. For example, an application needs a set of messages to communicate commands for call control (find route, analyze digits, disconnect call, etc.). The appropriate functions that generate these messages can be grouped into an ASE.

Common functions are the functions that all ASEs would use. Examples are security, remote operation, and association control functions. These functions are also usually grouped into appropriate ASEs and standardized as such. ROSE and ACSE are two examples of standardized ASEs that are widely used to support a variety of user-related ASEs. The common functions, which we can also call common ASEs, communicate with the layer beneath the application layer—the presentation layer.

An application uses an ASE through predefined parameters and, as a result, specific commands are issued. This can be represented by the example in Fig. 3.25. Here, application XYZ, which keeps track of calls (connected, disconnected, transferred, etc.) and is required to clear the call, is referenced as call 27. It uses a predefined parameter (parameter B) to invoke operations of a call control ASE at layer 7 of the communication model. This parameter carries a request to act: "disconnect call 27." The call control ASE refers to its records for call 27 to find other appropriate data and creates a message with the following information: call number, time of request, user identification, and a request to disconnect this call. This message (message G) is passed down to the next layer. Obviously, this is a made up example, but any application that uses ASEs behaves as described here. An application can use more than one ASE to achieve the results it needs. This depends on a particular application and the kind of services it needs from the available ASEs. Remember, though, that ASEs are created by either system developers (proprietary ASE) or standards developers. In either case the developers decide how to package functions into an ASE. The decisions are based on such factors as similarity of functions, optimization of ASE software development and debugging, and the size of potential ASEs. An important element in this decision is understanding the reason for a particular ASE development. For example, two standardized ASEs (ROSE and ACSE) are widely used since they were intended to be used in a modular fashion. Another ASE, TCAP, contains elements of both, in addition to other functionality. So, if TCAP is used, the ROSE ASE is not needed. This is because TCAP was specifically started as an SS7 part and, according to its designers, should contain all necessary transaction functionality.

In addition, when an ASE creates a message, this message traverses all layers of the protocol stack down to the physical layer, and then it is transmitted to the peer system. At the peer system the process is reversed—the message is received by layer 1 and delivered through the entire stack to layer 7. This is a “physical” manifestation of the communication. Logically, the peer ASEs communicate directly. This is illustrated in Fig. 3.26.

Comparison of SS7 and OSI protocol architectures. Some elements of SS7 clearly have roots in OSI concepts and standards. It would be useful to compare the SS7 protocol stack with the OSI model to determine their relationship.

If we compare the protocol architectures of SS7 and OSI, we would find some similarities. First, both are using the idea of layered protocols. They split the necessary communications functions into different levels (layers) and specify how interlevel communication is accomplished. Second, they both use the concept of peer-to-peer communications, which is the ability of peers to understand each other. Third, specific functions at some levels are the same in both protocol stacks, but not in all. This is because when the SS7 protocol development started, the OSI concept had not yet been fully developed. Also, while the SS7 developers were concerned with a specific protocol for a specific set of requirements, the OSI developers were creating a model that was intended to be applicable to a wide variety of future protocols. As the result, the SS7 functions that are simpler to understand,

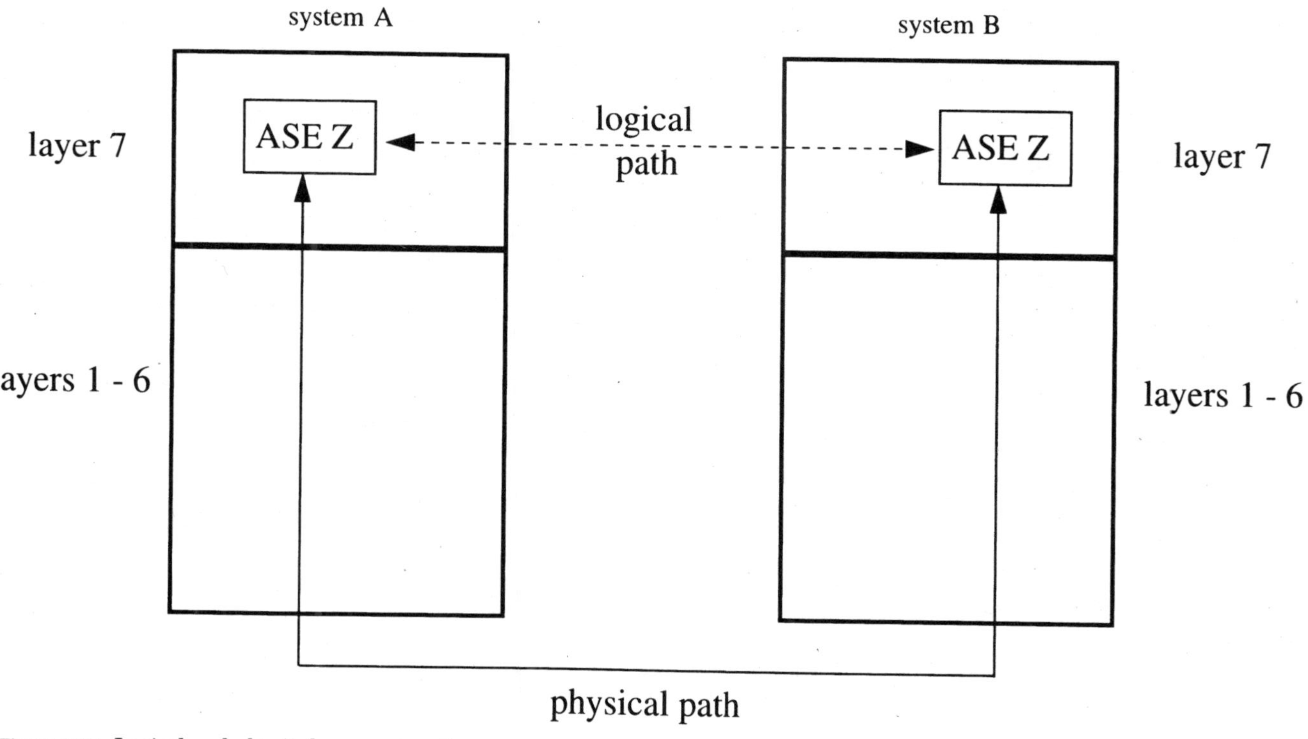

Figure 3.26 Logical and physical message paths.

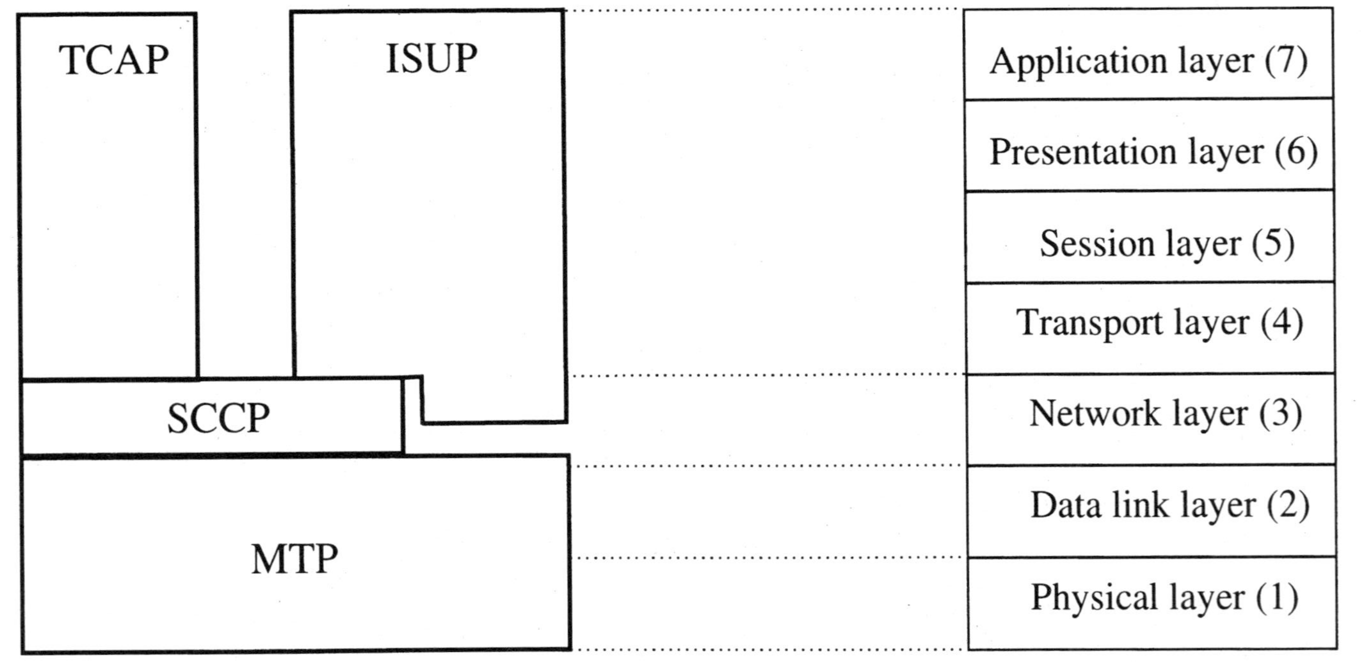

Figure 3.27 Comparison of SS7 and OSI reference models.

such as layers 1 and 2, are totally compliant with the OSI model. The functions at the higher levels are not always easy to fit into one or another OSI layer. A view of compatibility between the SS7 and OSI protocol architecture is shown in Fig. 3.27. Although TCAP and ISUP are shown to be spanning layers 4 through 7 (starting with layer 3 for ISUP), for the purposes of this book it is important to understand that TCAP and ISUP perform layer 7 functions.

3.3.5 IN standards

Standards for INs are developed in the form of models describing the concepts and a protocol that carries information between communicating systems (SSPs, SCPs). The protocol is an application layer protocol. Three aspects of IN standardization are discussed here: international standards, American national standards, and American de facto standards.

International standards. International IN standards are being developed by ITU-T, formerly CCITT. Some standards are already developed and others are either in the works or planned. The currently approved standards (the official name of an ITU-T standard is a *recommendation)* are discussed below.

Q.1200; *Q-Series Intelligent Network Recommendation Structure.* This recommendation lists all available ITU-T IN standards.

Q.1201; *Principles of Intelligent Network Architecture.* This recommendation identifies IN objectives, functional requirements, and architecture, including an IN conceptual model. The conceptual model consists of parallel planes:

- Service plane
- Global functional plane
- Distributed functional plane
- Physical plane

Figure 3.28 shows the conceptual model. It consists of four planes, each representing a different view of IN. This technique is used to separate service needs from functional and physical needs. Services are defined that are mapped to functions, which are then mapped to possible physical entities that can implement the functions.

The service plane represents the view of IN from the service point of view. Actually, it is not exactly a view of IN because IN specifics are not visible at this plane. Only services are visible, and they consist of some elemental service capabilities. For example, service "PCS" needs

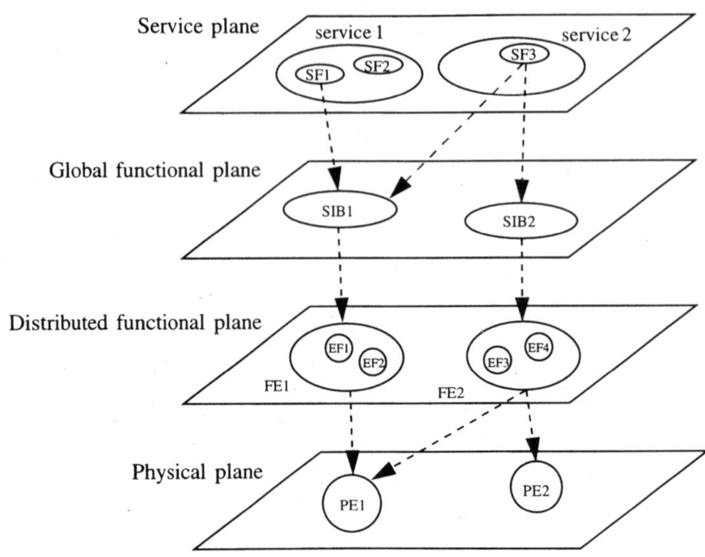

Figure 3.28 Simplified IN conceptual model.

call control, mobility management, radio resource management, etc. Service capabilities can actually be even more granular than in this example. In the model, service 1 consists of service features 1 and 2, while service 2 contains service feature 3.

The global functional plane identifies service-independent building blocks (SIBs) that are visible globally in a network. SIBs represent network capabilities that can support service capabilities of the service plane.

The distributed functional plane identifies specific functional groupings, called functional entities, that SIBs use to provide an element of service. These functional entities are composed of elementary functions. So, the progression from the service plane to the distributed functional plane is in terms of the scope of identified elements. At the top, service needs are identified that are gradually mapped to functions and then to actual equipment in the physical plane.

The physical plane identifies specific physical entities that can implement functions and the protocols that may be used between them. This is a view of a possible real network with IN capabilities.

Q.1202; *Intelligent Network Service Plane Architecture.* This recommendation describes the service plane of the IN conceptual model.

Q.1203; *Intelligent Network Global Functional Plane Architecture.* This recommendation describes the global functional plane of the IN conceptual model.

Q.1204; *Intelligent Network Distributed Functional Plane.* This recommendation describes the distributed functional plane of the IN conceptual model.

Q.1205; *Intelligent Network Physical Plane Architecture.* This recommendation describes the physical plane of the IN conceptual model.

Q.1208; *General Aspects of the Intelligent Network Application Protocol.* This recommendation describes the background and methodology used in the development of the IN application protocol.

Q.1211; *Introduction to Intelligent Network Capability Set 1.* The functions (capabilities) that IN will ultimately support cannot all be defined at the same time. In order not to delay the standardization of capabilities that the ITU-T membership agrees on at any moment, it was decided to produce IN standards in chunks called *capability sets*. The first one produced was capability set 1 (CS-1). The next one will be CS-2, and so on. Recommendation Q.1211 describes CS-1 capabilities (functions).

Q.1213; *Global Functional Plane for Intelligent Network CS-1.* This recommendation describes the global functional plane as it is used in CS-1.

Q.1214; *Distributed Functional Plane for Intelligent Network CS-1.* This recommendation describes all aspects of the distributed functional plane as it is used in CS-1.

Q.1215; *Physical Plane for Intelligent Network CS-1.* This recommendation describes the CS-1 physical plane.

Q.1218; *Interface Recommendations for Intelligent Network CS-1.* This recommendation defines an application protocol to support CS-1 functions. This protocol is the Intelligent Network Application Protocol (INAP).

Q.1290; *Glossary of Terms and Definitions of Intelligent Networks.*

Future IN capability sets will have numbering similar to CS-1. For example, CS-2 INAP will be described in Recommendation Q.1228; CS-3 INAP will be in Q.1238. The next set of IN capabilities is expected to be available in 1996—1997.

American standards. American national standards are being developed by subcommittee T1S1 of Committee T1 (Telecommunications One). T1S1 has worked very closely with ITU-T on the development of the international IN standards. T1S1, through the official U.S. channels, submitted many contributions to the ITU-T process and developed many concepts. Some U.S. companies were instrumental in the development of international standards, which largely explains why the American national standard will look like an international one when it is approved. Obviously, it is beneficial to have a national standard which is closely aligned with an international standard. But, since many countries have specific needs based on their history or regulations, national and international standards in many areas diverge. In the IN case this divergence is based on the U.S. environment. At this time, although specific differences are not clear, we can reasonably assume that most of the functional description will be based on the ITU-T standards, and the protocol will be based on the existing AIN protocol. The standard should be ready in 1996.

De facto standards. Standards defined and approved by ITU-T and the American National Standards Institute (ANSI) T1 are formal standards; that is, they were created by public committees specifically charged with such standards development. Many times the marketplace adopts as a standard a system or an approach that was created not by a formal standards committee, but by a member of a particular industry. MS-DOS and UNIX operating systems are examples of such systems that were developed, respectively, by Microsoft and AT&T and adopted by the industry as standards for which applications are written.

In the IN area such a de facto standard is AIN. AIN has been developed by Bellcore for its clients: Regional Bell Operating Companies, or RBOCs. It is based on IN concepts and is very similar to IN formal standards, but there are some differences. Most of these differences arise from the timing of the AIN and IN standards development. AIN development was happening faster, so that at the time when modeling issues were still being discussed within ITU-T, the AIN application layer protocol was being developed.

Bellcore's Technical Requirements (TR) and Generic Requirements (GR)[11] describe AIN and are being used for network planning and AIN deployment by major telecommunications companies. Although these

[11] TRs and GRs are basically the same things. GR is a new name for the type of document that used to be called TR. The process to produce a GR is a little different from the one to produce a TR, but the type of information they convey is the same.

TRs and GRs do not constitute formal standards, the fact that equipment is being manufactured and deployed according to them makes AIN a de facto standard. Currently, the following TRs and GRs are available:

GR-1129-CORE; *Advanced Intelligent Network (AIN) 0.2 Switch-Intelligent Peripheral Interface (IPI) Generic Requirements.* This GR addresses an interface between an AIN release 0.2 SSP and an IP.

GR-1280-CORE; *Advanced Intelligent Network (AIN) Service Control Point (SCP) Generic Requirements.* This GR addresses SCP requirements.

GR-1286-CORE; *Advanced Intelligent Network (AIN) Operations Systems (OS)-Service Control Point (SCP) Interface Generic Requirements.* This GR addresses an interface between an SCP and an OS.

GR-1298-CORE; *Advanced Intelligent Network (AIN) Switching Systems Generic Requirements.* This GR addresses requirements for SSP based on release 0.1, 0.2, and beyond.

GR-1299-CORE; *Advanced Intelligent Network (AIN) Switch-Service Control Point (SCP)/Adjunct Interface Generic Requirements.* This GR addresses an interface between an AIN release 0.1, 0.2, and beyond SSP and an SCP or an adjunct.

TR-NWT-001127; *Advanced Intelligent Network (AIN) Adjunct Generic Requirements.* This TR addresses requirements for an adjunct.[12]

TR-NWT-001254; *Advanced Intelligent Network (AIN) OS/Adjunct Interface Generic Requirements.* This TR addresses an interface between an OS and an adjunct.

TR-NWT-001284; *Advanced Intelligent Network (AIN) 0.1 Switching Systems Generic Requirements.* This TR addresses SSP requirements based on release 0.1.

TR-NWT-001285; *Advanced Intelligent Network (AIN) 0.1 Switch-Service Control Point (SCP) Application Protocol Interface Generic Requirements.* This TR addresses an interface between an AIN 0.1 SSP and an SCP.

---

12 An adjuct is an AIN network system similar to an SCP; it does not have an SS7 interface; it is usually collocated with an SSP.

3.3.6 AIN protocol

The AIN protocol has been designed based on the concepts discussed up to now. It is an application layer protocol created to support IN service, and it uses SS7 to exchange information between network elements. Information is exchanged in messages which are designed based on the needs of services.

The AIN protocol is being developed in phases, each new phase adding new functionality which is represented by new messages or new parameters in the existing messages. These phases are called *releases*. Two AIN releases were specified before the end of 1995: release 0.1 and release 0.2. Each of them identifies a set of functions, the later one obviously having a larger set. Which release or a set of functions is implemented in the equipment (switches, SCPs) depends on a particular manufacturer.

The goal of the AIN protocol is to support a variety of new applications, some of which are not defined today. Hence, it is being designed in the most flexible and accommodating way possible, so that additions can be made without changing existing functions. The protocol is described not in tabular form, as the previously designed protocols (X.25, DSS1, etc.), but in ASN.1—a descriptive language for the representation of structured information (ASN stands for Abstract Syntax Notation). It may be more difficult for an untrained eye to understand a protocol described in ASN.1 than the same protocol described in a tabular form, but remember that ASN.1 is just another language. Learning a few principles of this language allows a good understanding of what is represented. The advantages of using ASN.1 are based on its ability to represent information in a way which is easy for computers to understand, simplifying the process of migrating information into executable code. There are compilers available today that will take ASN.1 input and produce an executable code.[13] If all protocols in the world are described in ASN.1, various applications can cross-reference programs and data structures and simplify creation of new services. This would be similar to a computer system built out of many software components (programs) that cross-reference each other for operation and rely on the same set of values.

The message in Fig. 3.29 is presented as an example of AIN protocol elements. This is the Forward_Call message, which is an SCP response message directing a call to be forwarded.

This message defines an operation, the parameters used, and the errors that can be expected. The parameters are listed in the message

---

[13] This actually means that the ASN.1 description has to be developed with the optimization of the executable code in mind.

```
forwardCall      OPERATION
PARAMETER        SEQUENCE {
                 CallingPartyID                      OPTIONAL,
                 ChargeNumber                        OPTIONAL,
                 ChargePartyStationType              OPTIONAL,
                 CalledPartyID                       OPTIONAL,
                 OutpulseNumber                      OPTIONAL,
                 Tcm                                 OPTIONAL,
                 PrimaryTrunkGroup                   OPTIONAL,
                 AlternateTrunkGroup                 OPTIONAL,
                 SecondAlternateTrunkGroup           OPTIONAL,
                 Carrier                             OPTIONAL,
                 AlternateCarrier                    OPTIONAL,
                 SecondAlternateCarrier              OPTIONAL,
                 PassiveLegTreatment                 OPTIONAL,
                 PrimaryBillingIndicator             OPTIONAL,
                 AlternateBillingIndicator           OPTIONAL,
                 SecondAlternateBillingIndicator     OPTIONAL,
                 OverflowBillingIndicator            OPTIONAL,
                 AMAAlternateBillingNumber           OPTIONAL,
                 AMABusinessCustomerID               OPTIONAL,
                   SEQUENCE SIZE (1..2) OF
                 AMALineNumber                       OPTIONAL,
                 AMAslpID                            OPTIONAL,
                 SEQUENCE SIZE (1..5) OF
                   AMADigitsDialedWC                 OPTIONAL,
                 Amp1                                OPTIONAL,
                 Amp2                                OPTIONAL,
                 ServiceProviderID                   OPTIONAL,
                 ServiceContext                      OPTIONAL,
                 AMABillingFeature                   OPTIONAL,
                 AMASequenceNumber                   OPTIONAL,
                 RedirectingPartyID                  OPTIONAL,
                 RedirectionInformation              OPTIONAL,
                 CarrierUsage                        OPTIONAL,
                 ExtensionParameter                  OPTIONAL,
                 AMAServiceProviderID                OPTIONAL
                 }
ERRORS {         applicationError
                 }
::=private: 27137
```

Figure 3.29 Forward_Call message.

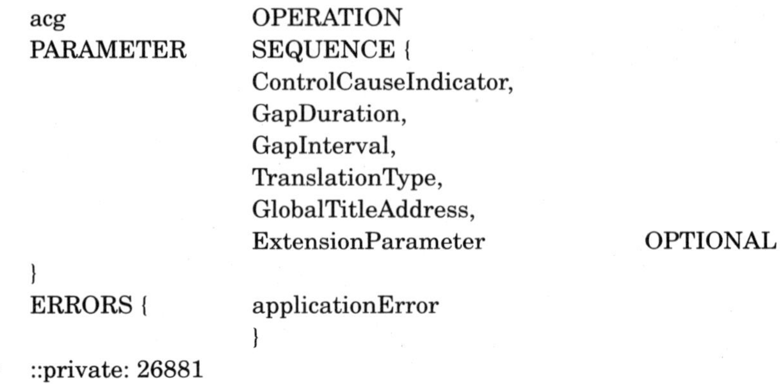

```
acg              OPERATION
PARAMETER        SEQUENCE {
                 ControlCauseIndicator,
                 GapDuration,
                 GapInterval,
                 TranslationType,
                 GlobalTitleAddress,
                 ExtensionParameter              OPTIONAL
}
ERRORS {         applicationError
                 }
::private: 26881
```

Figure 3.30 ACG message.

and each one is defined separately. In this particular message all parameters are optional. Other messages have mandatory and optional parameters, such as the ACG message (see Fig. 3.30). In this message parameters that are not explicitly optional are mandatory.

The AIN protocol messages are grouped into one of three categories: call related, non-call related, and abnormal. Call related messages carry information associated with a particular call. Examples of messages in this category are Info_Analyzed, O_Suspended, and Create_Call. Non-call related messages convey information which is not tied to any specific call. Examples of messages in this category are NCA_Data, Update, and Control_Request. Finally, abnormal messages,

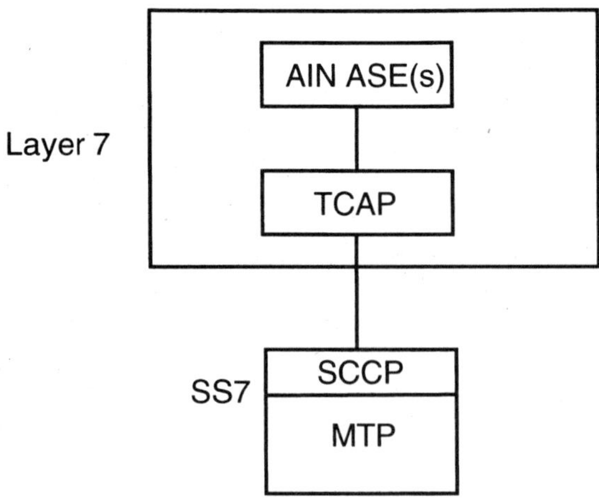

Figure 3.31 AIN protocol architecture.

specifically Application_Error, Report_Error, and Failure_Report, convey error-related information.

The AIN protocol architecture is shown in Fig. 3.31. It is quite simple from the modeling point of view: AIN ASEs use SS7, which is represented by its application layer part (TCAP), to transport information. The AIN protocol definition includes the identification of the TCAP usage. That is, the type of TCAP package is specified for each message.

AIN call flows. After looking at various pieces that make up IN operations, we can put them together and present some examples of what actually occurs when AIN call processing is used. These examples will show how triggers are set up and encountered, how an application message is formatted, and how SS7 is used to carry this message.

Service example 1. A subscriber requests that long distance calls from his home be barred Monday through Friday between 3 and 6 p.m. (presumably so that a child cannot spend money on long distance calls while adults are not at home). The implementation is as follows:

1. A call is originated from the subscriber's home on Tuesday at 4:10 p.m. The digits dialed are 415-731-9876 (see Fig. 3.32).
2. The SSP receives the dialed digits and begins digit analysis. The SSP performs this task in the Collect_Information state. In this state a trigger detection point is encountered—Info_Collected—with the trigger Off-Hook_Delay set for this subscriber. This trigger results in a message being formulated to the SCP (see Fig. 3.33).
3. The message to the SCP is formulated based on the information available to the SSP and according to the rules for creation of SS7 TCAP messages. Figure 3.34 shows how this message—Info_Collected—is constructed. The actual message has many parameters that are not specific to our example. Only relevant information is shown here. The package type of this message is Query with originating transaction ID = X. The message contains pertinent information about the caller (calling info) and the destination (called info). The message is sent on the CCS network with appropriate routing information.
4. An STP receives the message and finds that global title translation has to be performed to route this message to the correct destination. It changes the destination address from a global title to a physical point code of the SCP. The application-level information remains untouched (see Fig. 3.35).
5. The SCP receives the Info_Collected message and begins an analysis. It finds that the record for this particular subscriber (based on the calling number DN) indicates that no long distance calls are allowed at this partic-

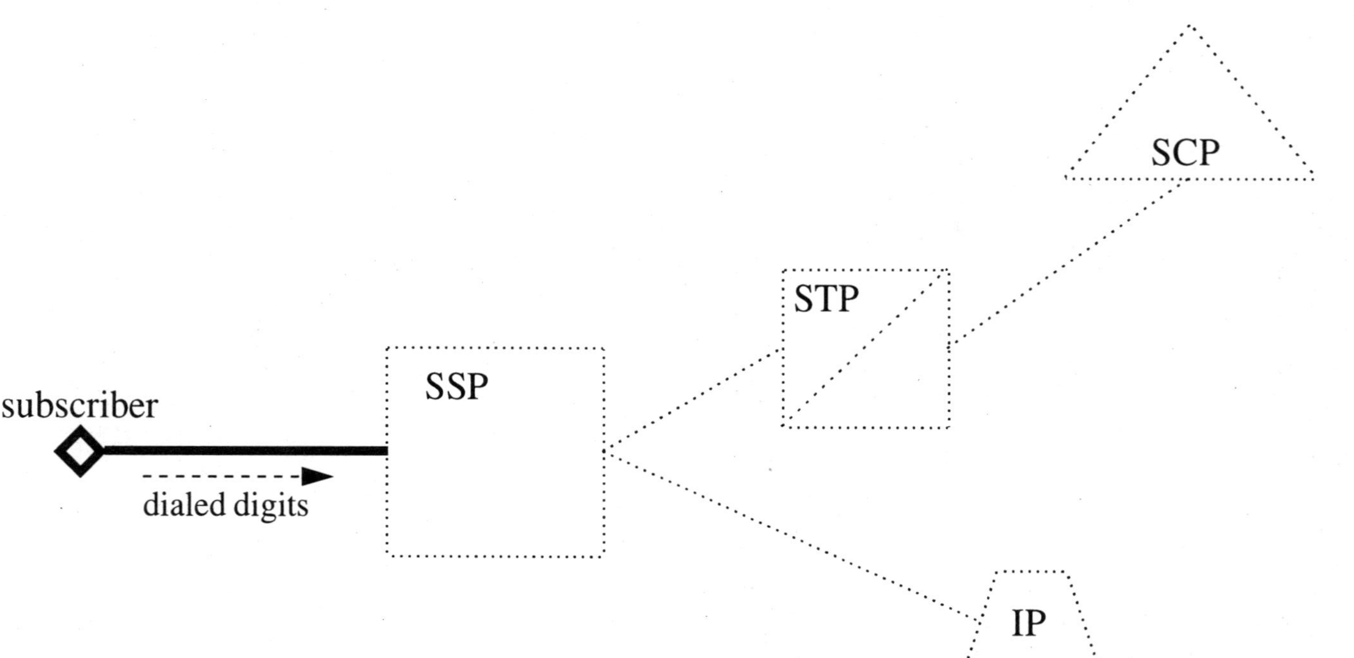

Figure 3.32 Call origination.

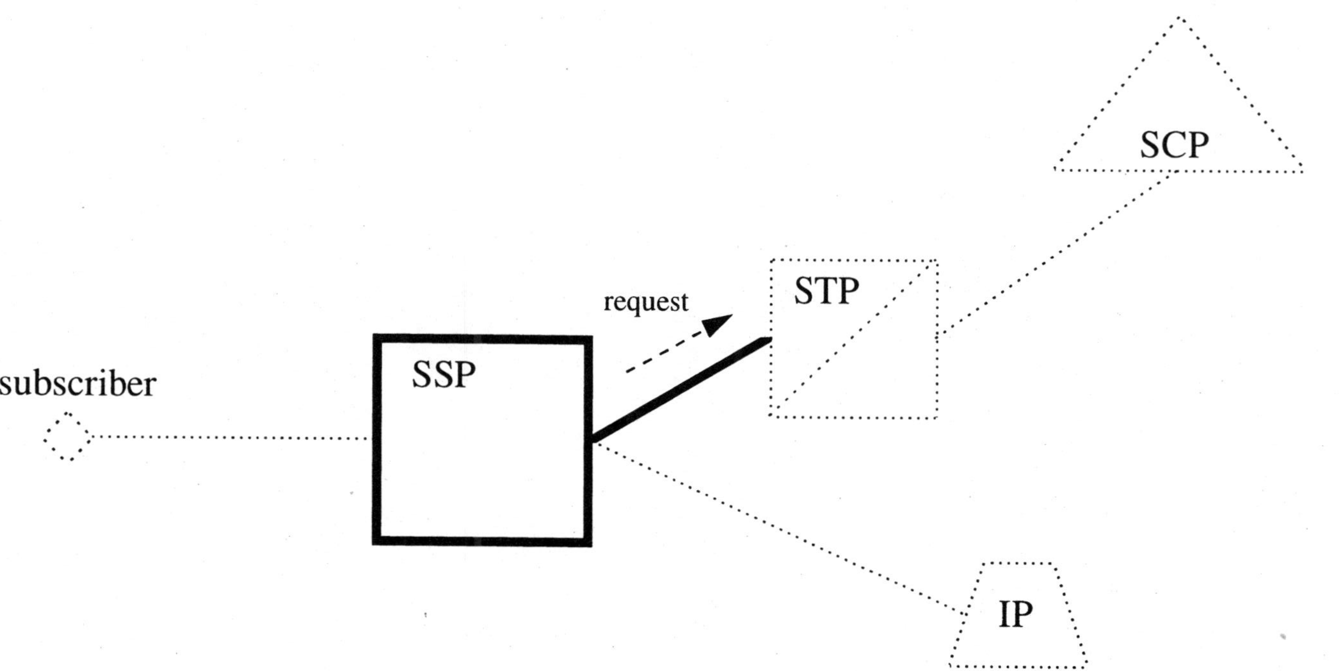

Figure 3.33 Request.

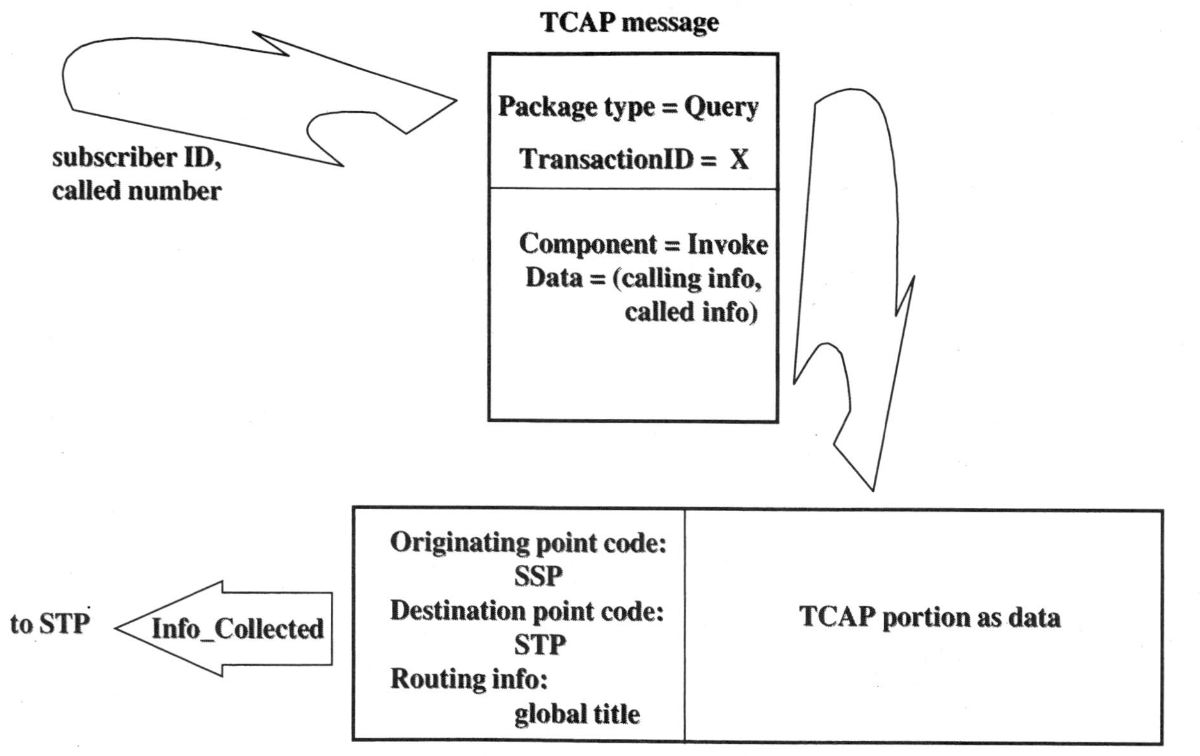

Figure 3.34 Info_Collected message construction.

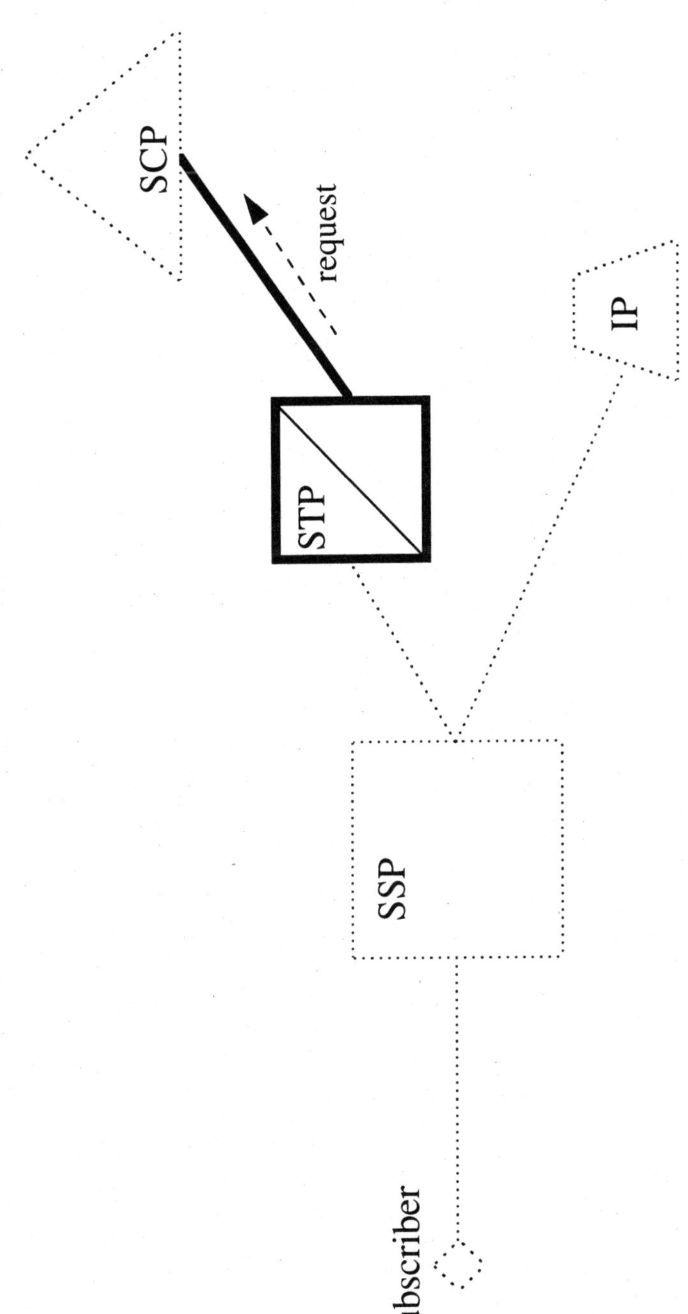

Figure 3.35 STP actions.

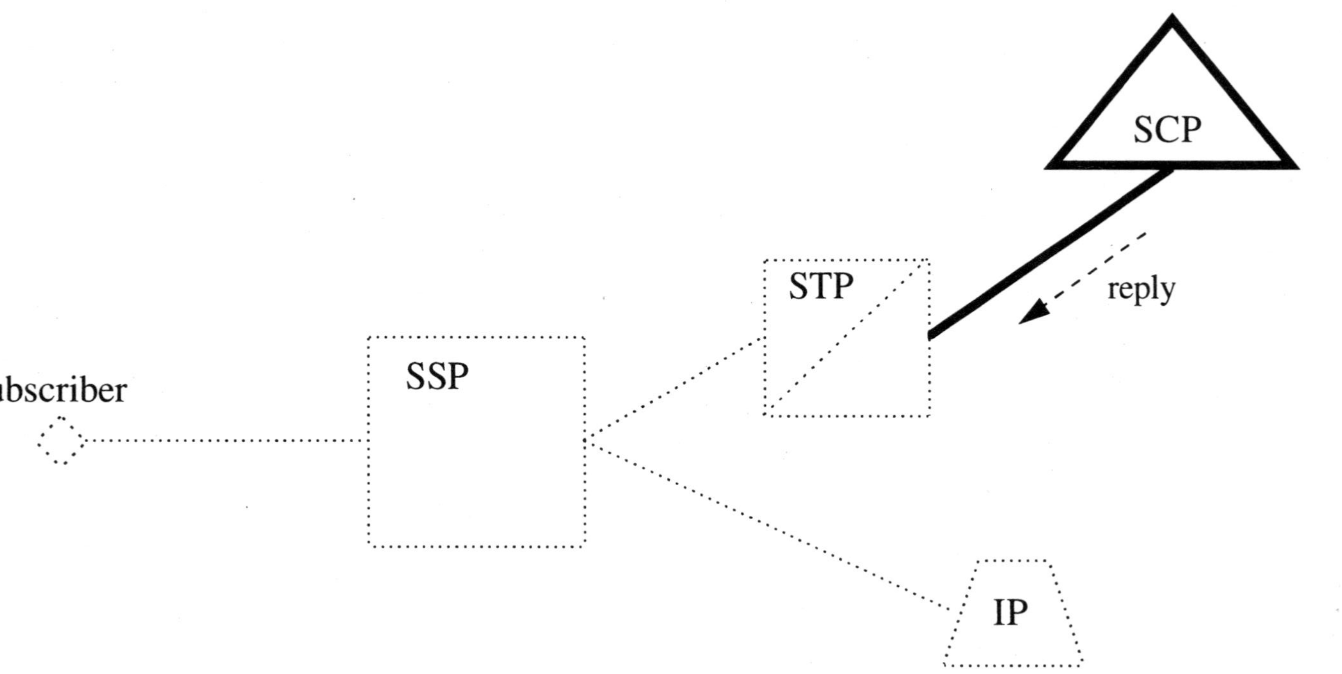

Figure 3.36 SCP actions.

ular time. After determining that the dialed digits correspond to a long distance call, it formulates a response to the SSP and sends it in a Send_To_Resource message (see Fig.3.36 on p. 68). This message is constructed according to the principles used in step 3. Figure 3.37 shows information carried in the message (only information essential to the service example is shown). The message package type is Response, because it is the last message for this transaction. The responding transaction ID is set to X (corresponding to the originating transaction ID in the incoming Info_Collected message). The data in the component portion of the message include information regarding the actions that the SSP would have to take in regard to this call.

6. The STP receives the message. No global title translation is performed because the point code of the SSP is included in the message since it was already known by the SCP sending this message. The message is routed to the SSP (see Fig. 3.38).

7. The Send_To_Resource message arrives at the SSP and contains the rejection for the originated call (in the form of DisconnectFlag parameter) and an address of a device that would issue a rejection announcement. This device is an IP. The SSP connects to the IP port indicated in the message and switches the subscriber line to the output of the IP for the announcement (see Fig. 3.39).

8. After the announcement has been played, the SSP disconnects the IP and clears all information related to the attempted call.

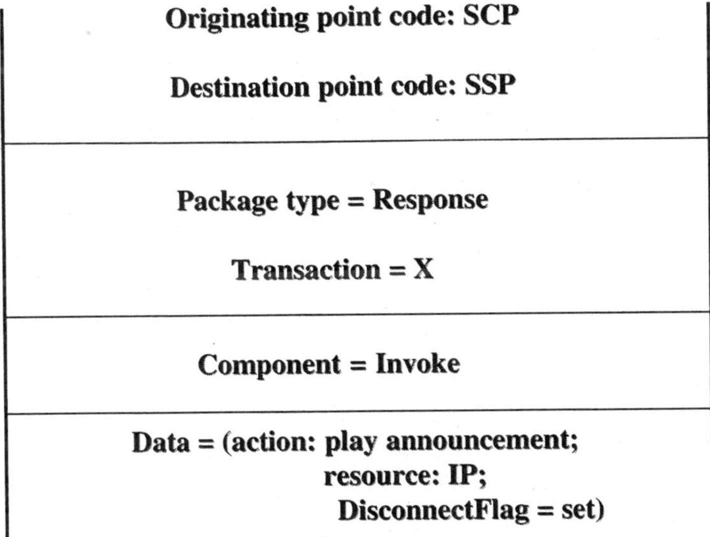

Figure 3.37 Sent_To_Resource message.

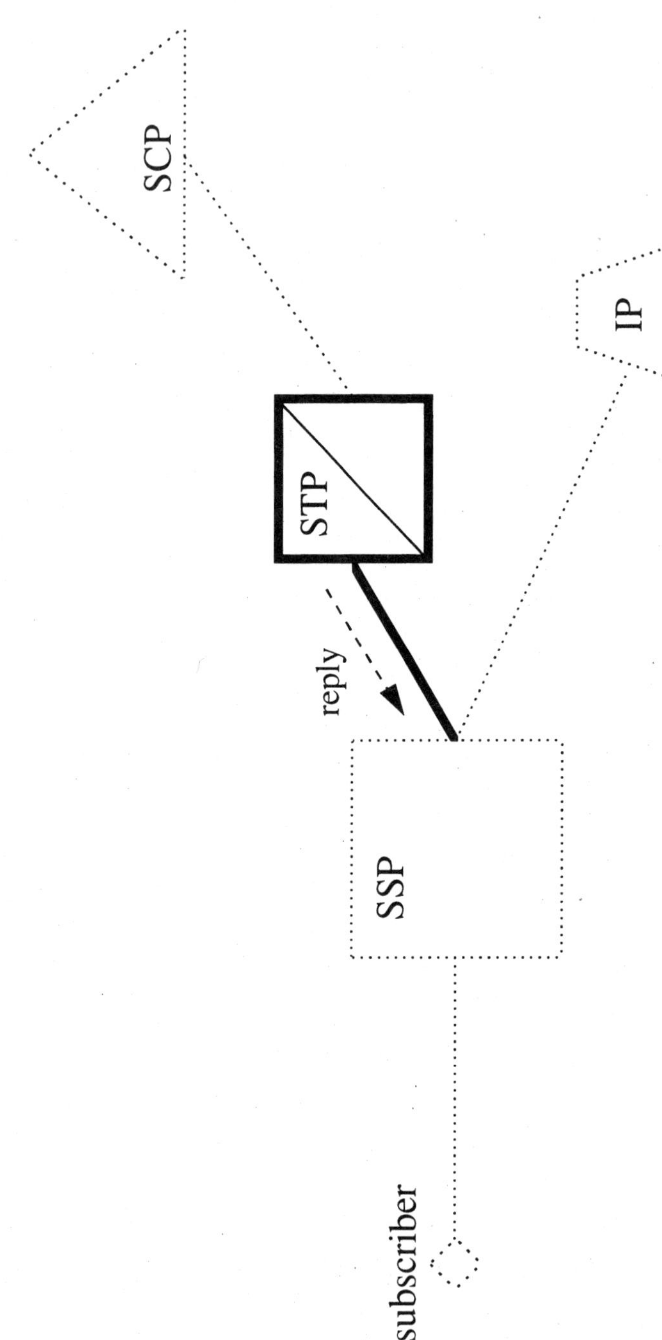

Figure 3.38 No GTT at the STP.

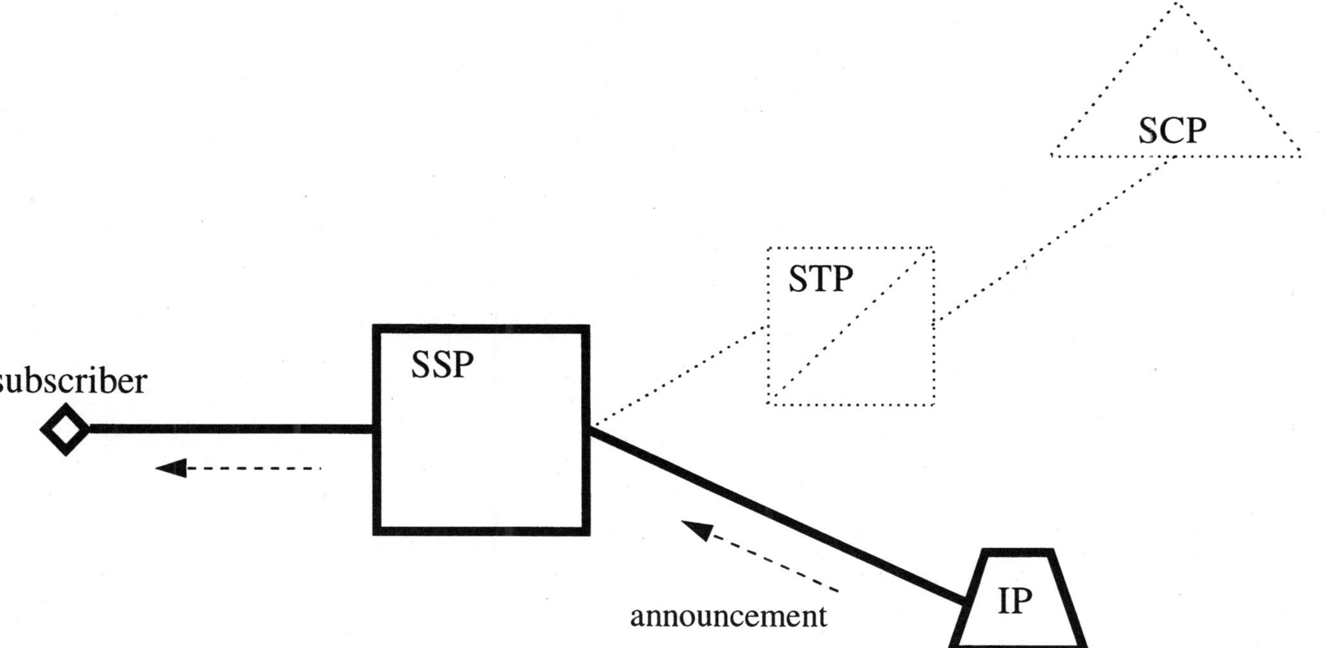

Figure 3.39 Announcement.

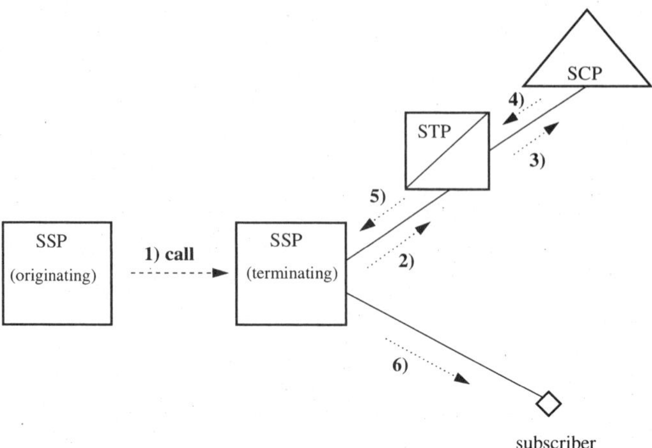

Figure 3.40 Example 2—terminating service.

Service example 2. The second example is based on the operation of the terminating trigger. In this example a user subscribes to a service which allows this user to receive and display information regarding another call which is coming in (we assume that the user is equipped with a display device). In this example all actions are described without a detail map of activities. Only a general figure is provided (Fig. 3.40). The implementation is as follows:

1. A call arrives for the subscriber at the terminating switch. The subscriber is using the phone, and when the Select_Facility state of the terminating call is reached, the T_Busy trigger is detected. This action generates a message to the SCP to be constructed—a T_Busy message. It contains, among others, the following parameters: busy cause, called party ID, called party station type, and calling party ID.
2. The T_Busy message is sent onto the SS7 network with the appropriate routing information.
3. The STP receives the message, performs a global title translation, and sends the modified message to the appropriate SCP.
4. The SCP receives the message, analyzes its contents, and sends the Offer_Call message to the SSP. This message contains, among others, the following parameters: calling party ID, display text, and service context. Depending on the service the customer subscribes to, the necessary information to be displayed at his or her device is included in the Display text parameter.
5. The STP relays the message to the appropriate SSP.
6. Display information is sent to the subscriber, who can now decide how to proceed with an incoming call.

Chapter

4

The Wireless World

The wireless world consists of networks, access to which is accomplished over the wireless interface using radio communications (see Fig. 4.1). The radio technology used for this purpose varies but accomplishes the same task: establishing a contact with a network. Wireless access has advantages and disadvantages that are the opposite of wireline access. The advantages are support for mobile users and quicker, simpler (in labor terms) installation and maintenance. The disadvantage of wireless access is the need to deal with greater interference than is experienced with hard wires.

4.1 Network Types

Two types of networks with wireless access are cellular networks and Personal Communication Services (PCS) networks. Both use radio to connect users to networks, and therefore, they are grouped together, using the network access type as the distinguishing characteristic. While cellular networks exist and serve customers, PCS networks are still on the drawing boards. Some PCS networks will become operational in 1996; they are being planned and designed right now. Thus, although PCS networks do not have actual customers that they serve, in terms of planning seamless networks, they are as real as cellular networks.

4.2 Cellular Networks

Cellular networks were built to serve mobile customers. From the very beginning their primary subscribers were people who required communications capabilities while they were moving. As the technol-

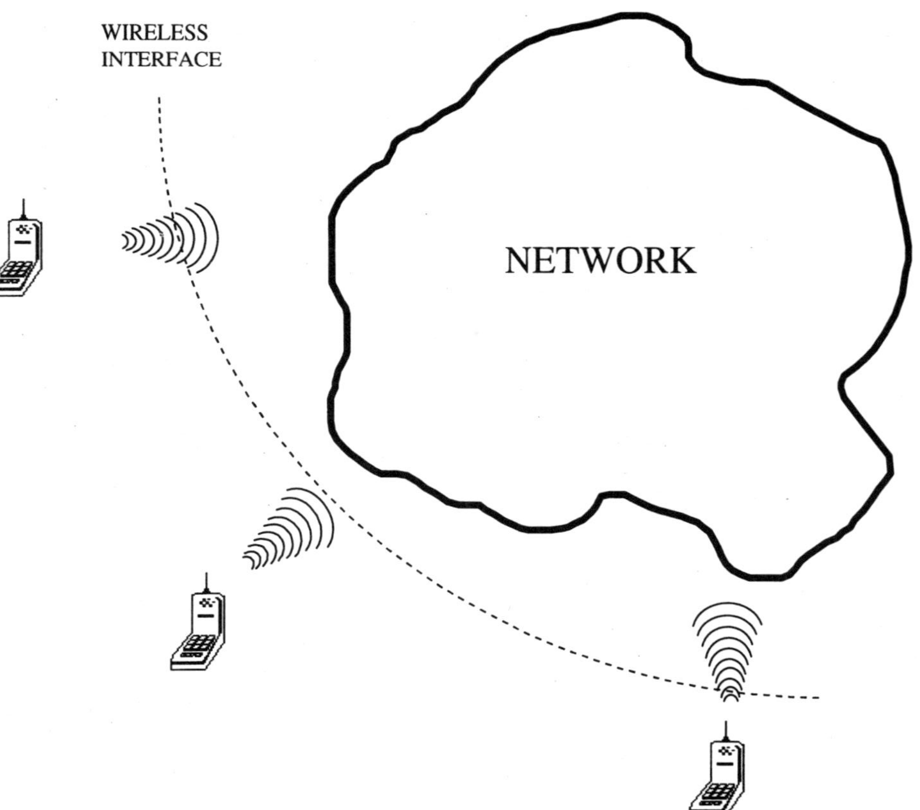

Figure 4.1 Wireless connectivity.

ogy improved, cellular customers began using phones while away from their cars. Right now, people frequently talk on a phone while they walk on the street, shop, or participate in business meetings (in which case everybody else can hear the ringing of an incoming call loud and clear). Cellular telephones are now much smaller than they used to be, their batteries have a longer use time, and the phones are easier to use, especially if they have digital displays.

4.2.1 Cellular services

Cellular networks provide various voice and data services, but voice services dominate. Their basic service is *call establishment,* as in the case of wireline networks. In the wireline world network availability (dial tone) and voice quality are excellent, but in the wireless world these aspects of service are more complicated because wireless subscribers move. It is more difficult to satisfy the needs of customers while they move than while their position is fixed.

Voice quality in cellular networks, although improved, has not yet reached the quality of wireline networks. Voice quality also suffers because users move around and are served by different radio ports while they talk. The process of switching users from one radio port to another due to changes in the users' position relative to radio ports is called a *handoff* (or handover), and it is a critical element of the service provided by a cellular network. The ability to hand calls off from one system to another has greatly improved the attractiveness of cellular service since it allows calls to be maintained while the user is moving. Without handoffs, the calls would be dropped if a subscriber moved far enough from the serving base station.

Another characteristic of a cellular environment is the user's movement from the area served by the user's service provider to an area served by a different service provider. This is called *roaming*. Roaming service allows calls to be established to and from a roaming subscriber. Roaming makes an overall cellular environment more attractive since it removes the service-only-in-a-designated-area restriction.

Call establishment, handoffs, and roaming are key basic services in the cellular environment, but there are also enhanced services similar to those provided in the wireline environment. Services for cellular networks are described in the Telecommunications Industry Association (TIA) interim standard IS-53 A; some of them are:

- Call forwarding
- Call waiting
- Three-way calling

- Remote feature control
- Voice messaging
- Message waiting notification
- Do not disturb
- Conference calling
- Calling number identification presentation
- Call transfer
- Flexible alerting

Some services are specific to the wireless environment, such as the following:

- *Short message service.* An ability to deliver a short message consisting of some critical data to a wireless handset.
- *Emergency services (911).* This is basically the same public safety service that the wireline networks provide, but it has a significant complication in its implementation: Mobile users are not attached to specific fixed locations, and therefore, it is difficult to identify the location of a 911 caller. The U.S. industry is currently working to solve this problem.
- *Voice privacy service.* An ability to safeguard voice communications.

Circuit- and packet-switched data services are also provided; they allow transfer of data at rates up to 14.4 kbps.

4.2.2 Architecture

Cellular and PCS networks are basically similar to wireline networks in the way they approach the needs of customers and provide capabilities for their services. However, wireless networks do have one important distinction—they have to deal with mobile (nonstationary) users and, therefore, have to know where these users are. These users are also moving across different networks and want to have an uninterrupted service, which also has to be supported by the networks. To support these basic needs, usually called *mobility management,* wireless networks contain some network elements that do not exist in wireline networks. Figure 4.2 shows a generic architecture of a cellular network. The elements of this architecture are described below.

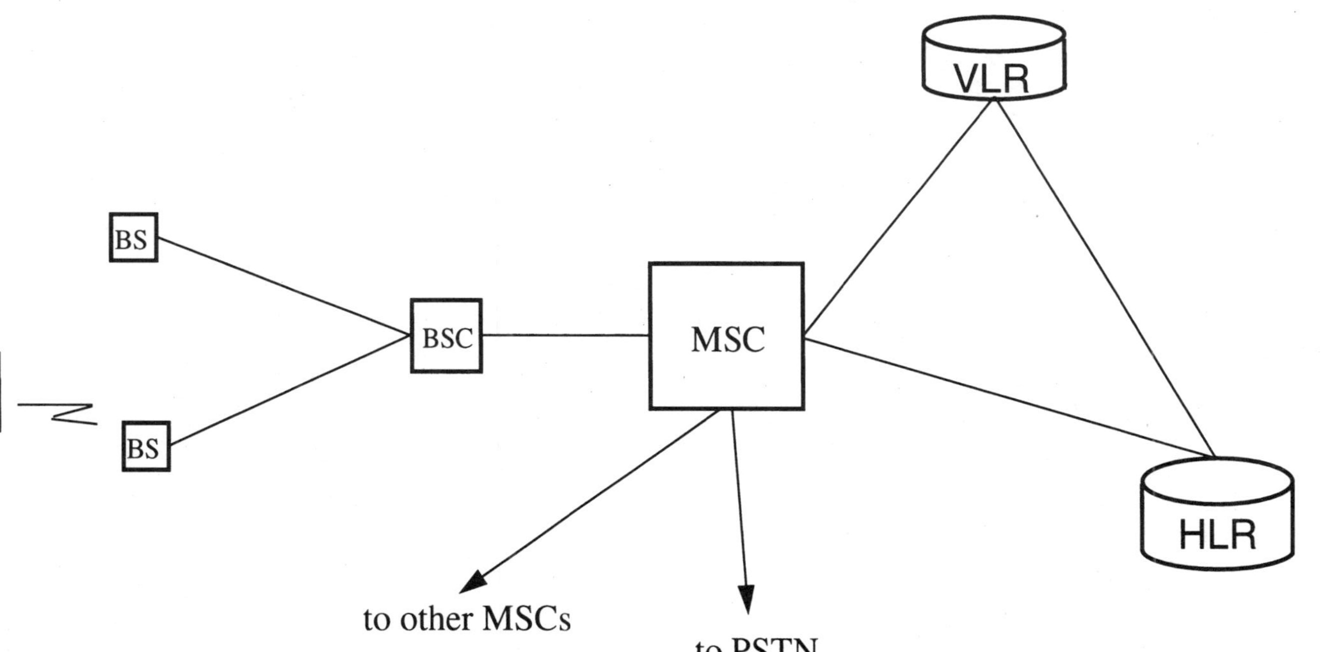

Figure 4.2 Cellular network architecture.

Mobile switching center (MSC).[1] The hub of the network; it performs switching functions and uses other network elements to find appropriate routes for calls. It is similar in functionality to a wireline network's switch but differs from it in its mobile-oriented functions. An MSC is connected to the Public Switched Telephone Network (PSTN)—the wireline network.

Home location register (HLR). A repository of all user information. It stores information about all users, such as names, billing addresses, subscriber profiles, and the current location of a user.

Visitor location register (VLR). A temporary repository of information for users who are away from their home area. VLR stores a subset of the HLR's information for a user who is roaming.

Base station (BS). A radio port that directly communicates with cellular subscribers over radio. It includes an antenna and electronics required to communicate with users and with the network.

Base station controller (BSC). An element that controls several base stations and communicates directly with an MSC. Whether BSCs are deployed in a network depends on the technology used by a particular service provider and the implementation of radio equipment by a particular manufacturer. It is possible not to have this element in a network, but BSCs often do exist, and that is why it is appropriate to show them in the architecture.

The cellular network architecture shown in Fig. 4.2 is functional. Physical implementations of cellular networks traditionally have been different. In most cellular networks a VLR is collocated with an MSC. This arrangement cuts down on the exchange of information between the MSC and VLR. Furthermore, an HLR, which can serve more than one MSC, is often collocated with an MSC. A typical cellular network is shown in Fig. 4.3. Although this type of implementation is viable, demand for wireless services will outgrow its existing performance level and require some of the functions to be handed-off by the MSC equipment. The HLR functionality would be the first candidate for such a change in the implementation of the wireless networks.

Interfaces. When a network consists of equipment produced by a single manufacturer, there is no need to pay attention to the interfaces between different network elements, except for engineering these

---

[1] Another name used for this equipment is Mobile Telephone Switching Office (MTSO).

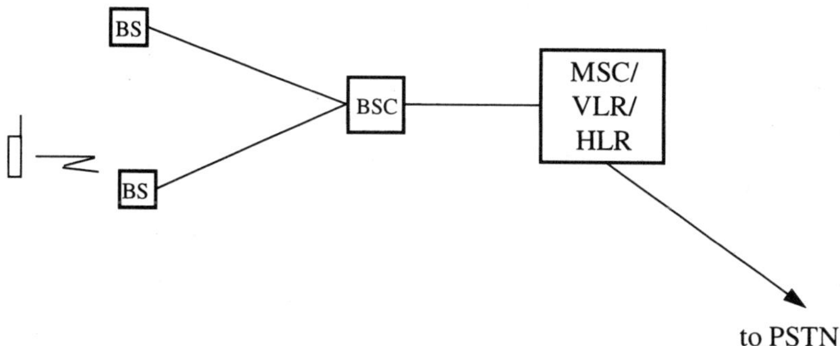

Figure 4.3 Traditional cellular network architecture.

interfaces for appropriate capacity and reliability. Since networks are usually built using equipment from more than one manufacturer, and, in addition, different networks have to interwork, many network interfaces are standardized. Before we discuss the cellular network interfaces, a standard network reference model should be presented (see Fig. 4.4).

This reference model has been defined by TR 45 of TIA and is based on the internationally standardized reference model for Public Land Mobile Networks (PLMN). It identifies all functional components of a cellular network and interfaces between them. Comparing this reference configuration with the typical cellular network architecture shows that they are quite similar. Some elements are present in the reference configuration and absent from our network architecture: Authentication Center (AC) and Equipment Identity Register (EIR).[2]

The AC is a functional network element that is responsible for privacy and authentication-related functions (e.g., encryption keys assignment). The AC's functions really depend on the level of privacy and authentication supported by the network. An AC can be managing information for more than one HLR in a network or for more than one network (as a clearing house), but currently the AC functions are incorporated within an HLR.

The EIR is a functional network element responsible for registering user equipment. Currently, this function is not used, but it is identified in the standards as a functional element which might be required.

[2] The latest variation of the TR 45 reference model contains two more elements: Short Message Entity (SME) and Message Center (MC). Both would be used to provide short message services—the ability to deliver a short message and display it on the mobile station (handset). It is not clear how these elements will be deployed in the cellular networks—as separate installations or as functions collocated within other elements (e.g., HLR, VLR).

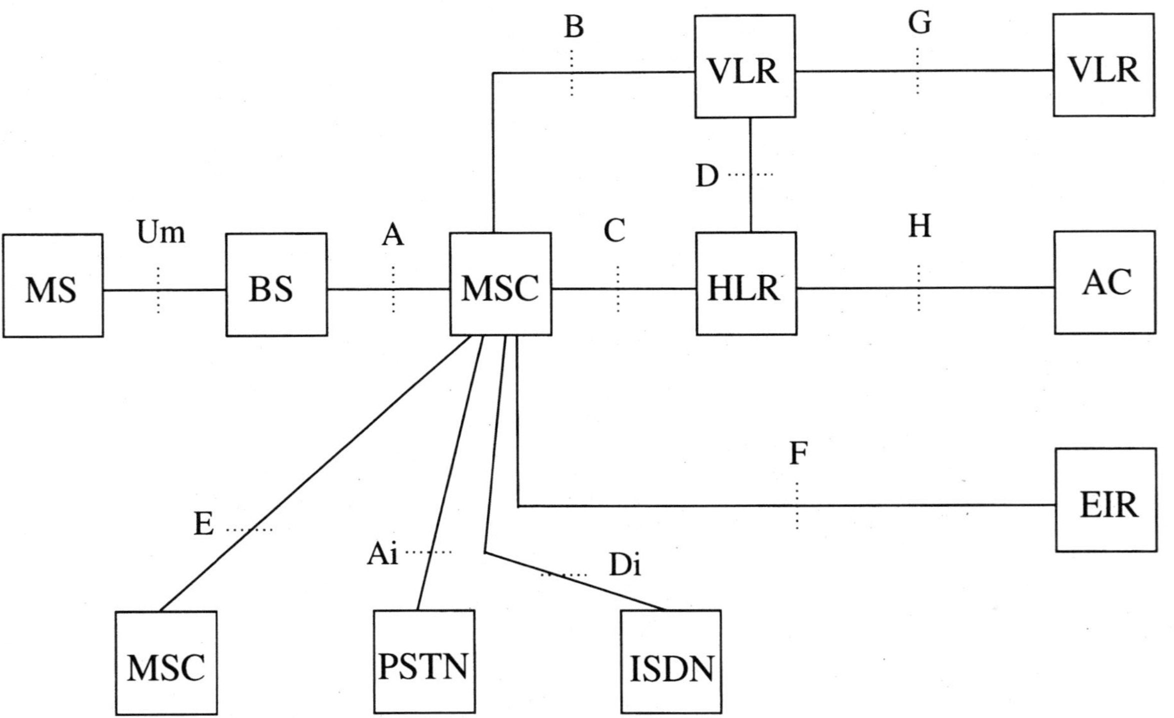

Figure 4.4 Cellular network reference model.

Another difference between the network architecture (see Fig. 4.2) and the reference model is in the designation of radio equipment. Although there is only a BS in the model, the network architecture has a BS and a BSC. Actual implementations of radio equipment by many manufacturers include both BSs and BSCs, and therefore, both are included in the network architecture. If BSCs are not used, some of their functions migrate to the MSC.

The reference model defines many interfaces, but we are only interested in the interfaces between physically remote network elements. Specifically, two interfaces are of interest: the air and the networking (intersystem) interfaces. They are of interest because equipment produced by different manufacturers meet at these interfaces. Handsets from different manufacturers have to talk to a number of different BSs, and MSCs and HLRs have to communicate with other MSCs and HLRs. Since, in most cases, MSCs and BSs deployed in the same area are from the same manufacturer, the interface between them has not been standardized, but this situation is changing.

Several types of standardized air interfaces are currently used in cellular networks. These are AMPS, IS-54/IS-136 Time Division Multiple Access (TDMA) and IS-95 Code Division Multiple Access (CDMA). AMPS is an analog air interface which is widely used in the United States. TDMA- and CDMA-based air interfaces are digital and are beginning to appear in cellular networks, TDMA being a more mature technology than CDMA.

Only one standard, fortunately, is used for the networking intersystem interfaces—IS-41. However, there are several different revisions of this standard.

IS-41 MAP. IS-41 Mobile Application Part (MAP) protocol standard has been developed by TIA TR 45 for the cellular industry. It defines functions and procedures for intersystem communications in wireless networks. IS-41 MAP protocol has primarily been defined to support handoffs, roaming, and some supplementary services, but it can be extended to support any appropriate requirements. IS-41 has been designed as an application layer protocol that uses available underlying transport capabilities. Its protocol architecture is shown in Fig. 4.5.

The first revision of IS-41, revision 0, was issued in 1988. The majority of cellular systems today are equipped either with revision A or revision B. Revision C has just been approved and work has started on revision D. The difference between revisions is functionality. The later the revision, the more functionality it possesses. For example, the main addition in revision B is the capability to use global title translation (GTT) for routing. The additions in revision C are more numerous; they include:

- Authentication and encryption
- Support for CDMA
- Border cell problem resolution
- Greater number of supported features

The current IS-41 revision which governs the development plans of manufacturers and service providers is revision C. It contains a large set of operations that can be used for a variety of functions. The philosophy behind IS-41 is based on reusability and extendibility. That is why existing protocols are used as transport mechanisms: Transaction Capabilities Application Part (TCAP), lower levels of Signaling System #7 (SS7), and X.25. The process of defining a new operation in IS-41 is as follows.

The operation is defined in terms of the information (i.e., parameters) it carries and the message is described in terms of a request and response. That is, a message requesting the operation (Request component) and a message responding to the request (Return Result component) are specified. A package type is assigned for all components. Finally, procedures associated with this operation are defined.

When the request is successful, a return result is sent back. When the request is not possible to grant, an error is returned. And if a TCAP message is not understood (e.g., incorrect formatting, etc.), a reject is issued.

Tables 4.1 and 4.2 show an example of an IS-41 operation as a request and a successful response. These two messages represent a

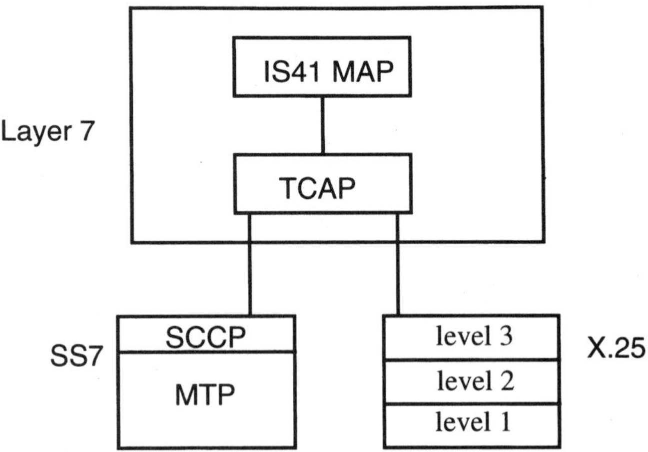

Figure 4.5 IS-41 protocol architecture.

Table 4.1 AuthenticationDirective Invoke Parameters

Parameters	Type*
Identifier	M
Length	M
ElectronicSerialNumber	M
MobileIdentificationNumber	M
AuthenticationAlgorithmVersion	O
AuthenticationResponseUniqueChallenge	O
CallHistoryCount	O
DenyAccess	O
LocationAreaID	O
RandomVariableSSD	O
RandomVariableUniqueChallenge	O
SenderIdentificationNumber	O
SharedSecretData	O
SSDNotShared	O
UpdateCount	O

*Indicates a mandatory (M) or optional (O) presence of the parameter.

request and a response for a successful authentication directive operation. If the request were formatted incorrectly (e.g., had an unknown identifier), a reject would have been issued. If the operation cannot be performed (e.g., the electronic serial number is incorrect), an error return is issued.

4.3 PCS Networks

What is different about PCS networks? What is the distinction between cellular and PCS? There may be many answers to these questions, depending on the point of view. Many people would base their answers on differences in services. However, services are not likely to be a distinguishing characteristic, at least not in the beginning of PCS networks. Up to now one clear distinction has surfaced: spectrum. Cellular networks operate in the 800-MHz band, and PCS net-

Table 4.2 AuthenticationDirective Return Result Parameters

Parameters	Type
Identifier	M
Length	M
CallHistoryCount	O

works will operate in the 1800-MHz band. This shift in frequency makes a difference in radio equipment design requirements and performance, but within the networks this difference is invisible.

4.3.1 Services

Although it is envisioned that PCS services will be greatly different from those provided by the cellular networks, the first PCS services seem to be very similar to them, if not the same. Therefore, it is hard to identify the services that can only be provided in the PCS environment. The basic services, such as call establishment, handoffs, and roaming, are functionally the same as in cellular networks.

At this time, the most comprehensive public view of PCS services is documented in the IS-104 interim standard produced by TIA. It has a list of 28 supplementary services plus data services. Many traditional services are listed there, such as call forwarding, call waiting, message waiting notification, and others. These are the same services that wireline and cellular networks provide. There are also some services that do not appear on the list of services defined for cellular networks (see TIA interim standard IS-53 A), specifically, they are:

- *Barring of incoming calls.* An ability to indicate to a network that certain calls should not be delivered
- *Closed user group.* An ability to identify a group of users working together
- *Completion of calls to busy subscribers.* An ability to complete a call to a subscriber who was busy during the original call completion attempt but is available later to receive a call.
- *Automatic reverse charging.* A feature that allows charging for services to be directed to a predefined number.

This is not an exhaustive list of specific PCS services, but the important point is that they are not, in their current state, principally different from cellular services. Regardless of the differences that exist between the two standards documents describing PCS and cellular services, the type of services to be supported in both is the same. There are no earth-shattering PCS services that are specified. There could be services related to personal mobility, but they could also be supported by the cellular networks.

It is possible that in the near future the cellular and PCS services described in IS-53 A and IS-104 will be merged. Then, from the point of view of standards, there would not be any differences between PCS and cellular environments. Actual implementation differences may appear when PCS networks become operational. It is possible that the new digital technology that will be used in PCS networks will allow the implementation of services faster, cheaper, and more efficient, but this becomes more a question of competition than of the relative abilities of cellular and PCS networks. Since services differentiate service providers, it will be interesting to watch how they are implemented in existing and new networks.

4.3.2 Architecture

The architecture of PCS networks is the same as that of cellular networks. PCS networks address the same kind of users with similar equipment in the same environment. The same types of network elements are required as in the cellular networks:

- MSC
- HLR
- VLR
- BS
- BSC
- AC

While functional network elements are the same, the actual equipment that will be deployed in the PCS networks will be different. Even if in the beginning, in order to roll out services quickly, some of the existing cellular-based equipment may be used (such as that supporting AMPS), the goal of PCS networks is to build modern digital networks that can provide higher quality and more sophisticated services. Because of this we will treat PCS networks as digital. However, deciding which particular digital technology will be used is not easy, and the selection is quite wide.

4.3.3 PCS technology

When Europeans began building digital wireless networks, a single standard emerged—Global System Mobile, or GSM. In the United States the situation is different. Several digital technologies are available and there is no single standard. Potential PCS network providers have many technological choices. Each of the PCS license holders will have to decide which technology to use, and this decision will have to be based on many factors: economics of a particular technology, its general availability, quality, support of enhanced services, and other factors. Here is the list of these technologies:

- TDMA
- CDMA
- GSM
- Omnipoint
- Personal Access Communications Systems (PACS)
- Digital European Cordless Telecommunications (DECT)

At the time of this writing, GSM is widely deployed internationally (but not much in the United States), TDMA is being introduced, and the rest are getting ready for competition. Because the competition for the place in the sun is tough, companies supporting particular preferences in technology make claims about the benefits of one selection and the disadvantages of the selections supported by the competition. The claims are in the areas of capacity, coverage, capital needed to build a network, security, etc. These claims and arguments will be settled when a particular technology is deployed. We will learn who the winners are later. Right now, we should treat all of the above as contenders and assume that any one of these technologies represents a workable solution for a PCS network. The differences in technologies have two aspects: (1) different technique for using the spectrum (time division versus code division) and (2) different solutions using the same technique (e.g., GSM versus PACS). Note that these technologies are different radio technologies, that is, they provide different ways of wireless network access. The basic network architecture required to support these access variations is the same.

TDMA. TDMA technology is not new. Time division is used in many different applications of digital transmission. Its key is the procedure of dividing a unit of time (e.g., a second) into time slices and treating each slice (called a time slot) as a separate transmission channel (see Fig. 4.6). Each channel (time slot) is used by a specific transmitter to transmit information, and when it is received, the receiver knows who transmitted it by identifying the channel number. Every transmission frame includes time slots for each channel. A transmitter stores information in the appropriate time slot, and a receiver extracts this information from each time slot (channel).

This is the basic distinguishing characteristic of TDMA technology, but when people in the United States say "TDMA," they usually imply a specific way of using the TDMA technique, which is standardized by TIA (see TIA IS-54 and IS-136).

CDMA. CDMA is also not a new technology. It has been used in military applications for quite a long time, and it is only now making its way into

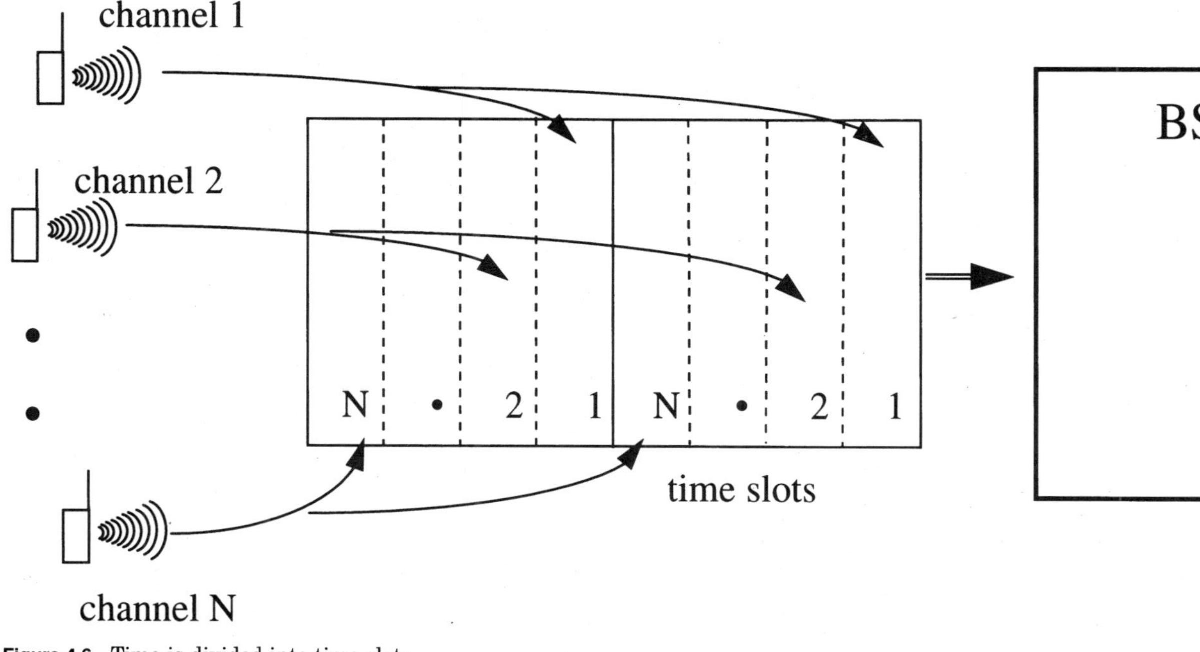

Figure 4.6 Time is divided into time slots.

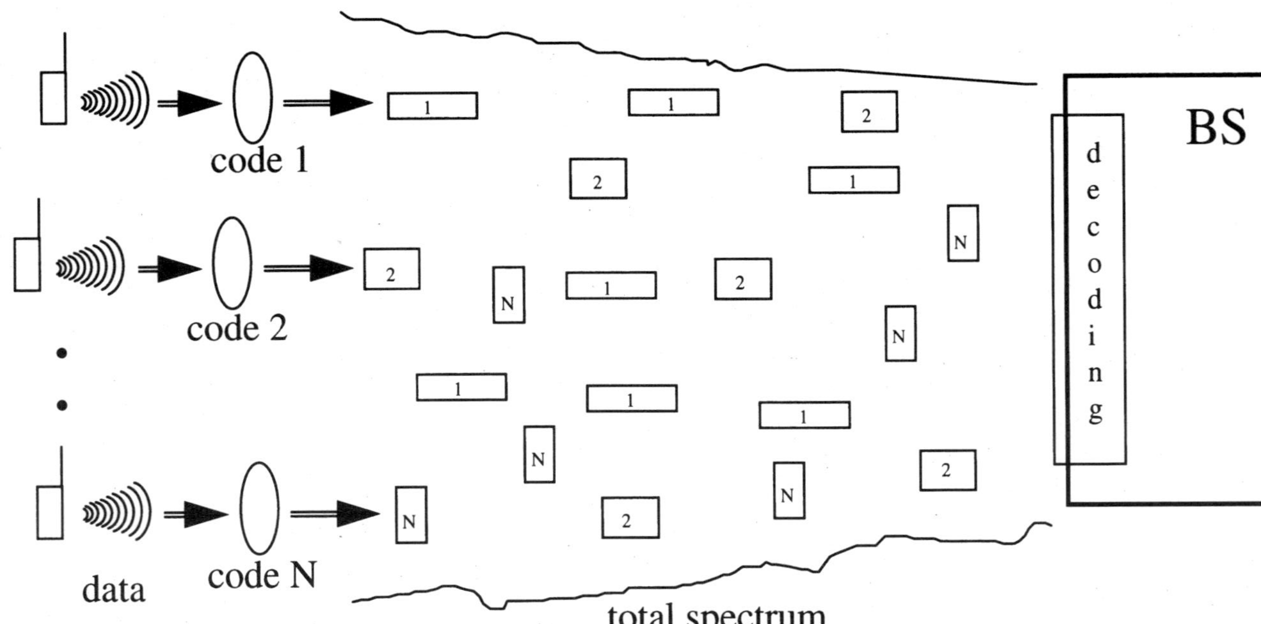

Figure 4.7 Codes are used in CDMA to spread spectrum.

the consumer market. In CDMA, information is transmitted using a specific code assigned to each user and applied to all transmitted information in order to spread data over a wider bandwidth. A receiver distinguishes the transmitter by knowing its code (see Fig. 4.7).

CDMA systems have two important characteristics that affect their operations. The first is the need to control the transmit power of a CDMA handset. When the handset is closer to the BS, its transmit power is usually decreased, and when it is far from the BS, it is usually increased. This is referred to as a *near-far* problem, and its solution is key to deploying commercial CDMA networks.

The second characteristic is a unique form of handoff called a *soft* handoff. In a typical hard handoff situation, used by TDMA systems, when a new channel is acquired, the old one is dropped. With a soft handoff two (or more) channels are active and, for the up-link (the information flow from a handset to a network), the network selects the data from the channel that is identified as being of better quality. For the down-link (the information flow from a network to a handset), the handset combines signals from the active channels to derive data of higher quality than from a single channel.

GSM. GSM[3] is a TDMA system. It uses exactly the same principles as the TDMA system standardized in the United States. However, GSM was created in Europe and standardized as a system; that is, all interfaces where different manufacturers' equipment can meet are standardized. Although transmission principles of GSM are the same as IS-54 TDMA, the call control and mobility management procedures are different. Data structures used for the system operations are also different (see Chap. 10 for more details).

Formally, GSM refers to the European standard applicable for the 800-MHz band, and DCS 1800 is the European version of this technology for the 1800-MHz band. Its adaptation in the United States is called DCS 1900. In this book GSM refers to the technology unless a specific European or American standard is addressed.

Omnipoint. This radio access technology has been developed by a company having the same name. Although it is being standardized and has a formal standard number, many people still refer to it by the name of its inventor—Omnipoint. This technology constitutes a mix of TDMA and CDMA methods and promises the best of both approaches.

---

[3] Its original title was Groupe Speciale Mobile, which is in French. As GSM gained popularity, its name changed, while the letters by which people refer to this system remained the same.

PACS. PACS is also a TDMA system. It was invented by Bellcore and geared for low-power applications, such as pedestrian traffic or for a wireless PBX environment. The latest trials of this technology also show that subscribers moving at significant speed (50 mph and possibly higher) can be successfully served by PACS systems.

DECT. DECT is, as its name implies, another European standard. DECT is a TDMA system designed for low-power applications. Some of the signaling techniques it uses are derived from GSM.

4.3.4 Standards

A description of standards for PCS can fill a thick book. There are many standards, since there are many potential technologies, and there are many standards committees that develop these standards. A newcomer to this field could get lost very quickly. Therefore, we will identify who is doing what. In addition to explaining what is going on in the PCS standardization, this information will be helpful when we discuss interoperability issues and how standards can help.

Understanding PCS standards requires understanding the different forces that are affecting them. First, there are a variety of radio technologies that can be used for PCS. Companies advancing specific technologies are making sure that any standards work does not exclude these technologies and thereby make them less competitive. Second, all existing standards bodies want to participate in the decisions (competition again). And third, there is competition between supporters of cellular and PCS environments. This is quite interesting because while this competition exists, some important cellular standards are also becoming PCS standards, but this third competitive force may be disappearing because many companies are getting involved with both cellular and PCS markets.

Two domestic standards committees accredited by ANSI are involved with PCS standards: T1 (Telecommunications one) and TIA. T1 is traditionally the committee where standards for wireline networks have been developed since the AT&T divestiture. T1 is involved with SS7, Integrated Services Digital Network (ISDN), Broadband, and other standards. TIA is the committee which, among other things, developed cellular standards. Both committees participate in the PCS standards development.

TIA. TR 45 of TIA is the committee that developed cellular standards, including IS-41. A new committee, TR 46, was created relatively recently to handle PCS standards. This committee is supposed to produce all PCS standards: air interfaces, network interfaces, interworking, and other standards. In practical terms, TR 46 develops A-interface (for the A reference point identified in the reference configuration) standards based on SS7 (GSM-driven) and ISDN (IS-41-driven) network access mechanisms. For example, an IS-651 standard has been developed which specifies an SS7-based A-interface, which is basically a U.S. GSM standard that also professes to support CDMA technology. A comparable ISDN-based standard is almost ready.

TR 46 is beginning to investigate other network access mechanisms, such as frame relay and Asynchronous Transfer Mode (ATM). This work has just started, and it will be some time before a clear view of the industry emerges. Another important work item is interworking between IS-41 MAP and GSM MAP—two incompatible networking protocols. This work will benefit seamless networks.

While TR 46 is also nominally responsible for the development of a networking protocol for PCS, in reality this protocol is IS-41, developed by TR 45, which continues its enhancements to suit other requirements. Regardless of what will happen in TR 46 regarding IS-41, its support for PCS is included in the work that is underway right now, including work for wireless intelligent networks (WINs). Clearly, the same networking protocol for both cellular and PCS environments would make interoperability between them a lot simpler.

It is also possible that in the near future TR 46 will be left only with the GSM-related work, while all other tasks may migrate to TR 45. Then, TR 45 would formally be responsible for the development of the IS-41 protocol and services for PCS, as well as for cellular applications. This will further blur distinctions between PCS and cellular networks.

T1 Committee. In T1 the overall responsibility for PCS standards has been assigned to T1P1, one of the technical subcommittees (TSCs). It is specifically responsible for all air interfaces and definitions of PCS services. T1S1 TSC is also involved with PCS standards, specifically with the development of protocols that are required to support PCS. T1S1 develops and maintains (e.g., assigns values for parameters) some of these required protocols, such as ISDN and SS7. Other protocols will have to be developed. There is actually a formal methodology defined for the development of any communications protocol. The methodology has been standardized by ITU-T and is used by many regional standards committees around the world.[4] This methodology consists of three stages:

---

[4] Regional standards committees are either responsible for a specific country's standards (e.g., American National Standards Institute, or ANSI, in the United States) or for a specific region (e.g., European Telecommunications Standards Institute, or ETSI, in Europe).

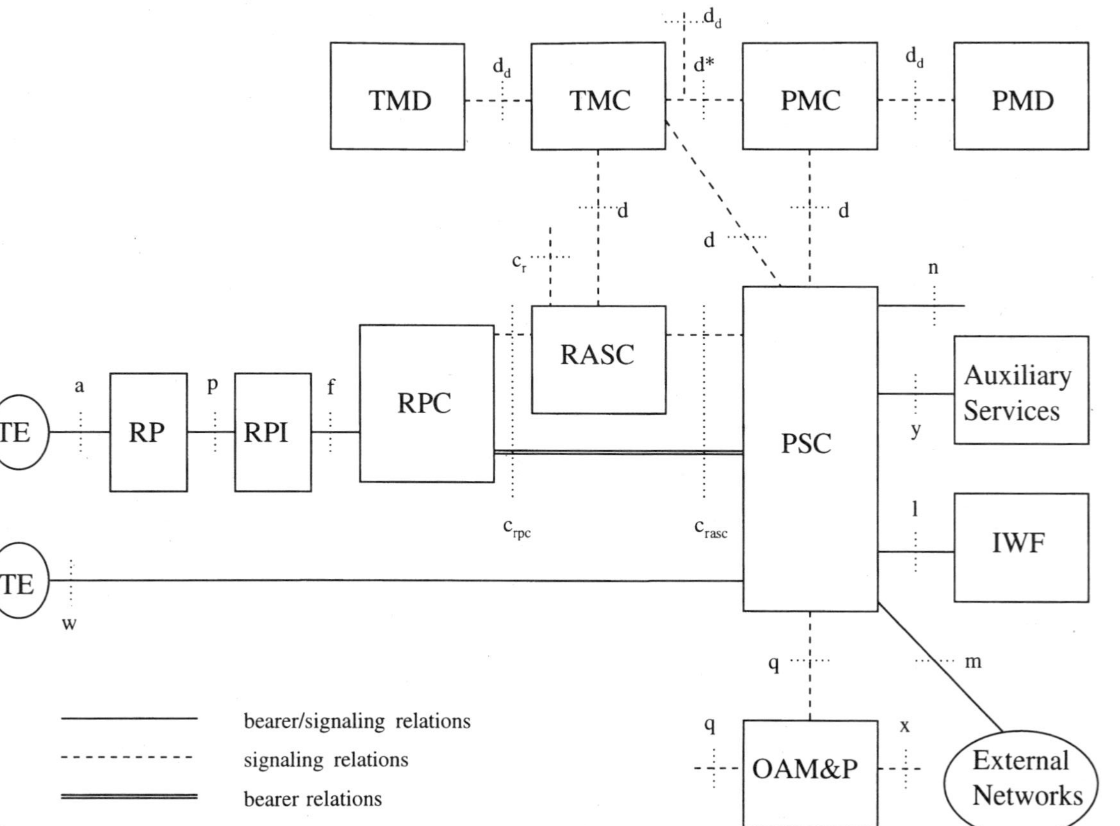

Figure 4.8 T1P1 reference model.

- *Stage 1.* The required services are described from the point of view of users, that is, services are described as a user would invoke them.
- *Stage 2.* The required services are described from the point of view of the networks that would provide them. Functional dependencies of different functional network elements are described. In short, at this stage functional requirements for networks are defined.
- *Stage 3.* The actual protocols required to support the services are developed. These protocols would be based on the requirements of stage 2, which were originally based on the needs of users defined in stage 1.

This methodology has been used by ITU-T and T1 in the development of ISDN and other platforms that came after ISDN, such as the intelligent network (IN). It was only natural for T1 to adopt the same process for the development of PCS standards. The result is that T1P1 has responsibility for stage 1 and stage 2 development, and T1S1, with its great expertise in protocol development, has responsibility for stage 3 development. But this is only a beginning. In addition to different stages, responsibilities for different interfaces are also split. Before we identify these responsibilities, a discussion of the PCS reference model is in order.

One might say that we already talked about a reference model and should have a clear understanding of it. That is true from the TIA perspective, but in T1 a different reference model has been developed (see Fig. 4.8).

The T1P1 reference model is functional, as is the TIA reference model, but it attempts to be more generic. This approach would presumably allow a greater variety of network implementations. The result is that terms such as VLR and HLR do not exist at all. Instead, functional capabilities are identified under new names. They are described in the following text:

- *PSC.* PCS switching center; it is the network entity that performs switching functions, e.g., an MSC.
- *TMD.* Terminal mobility data-store; it provides database functions for terminal mobility services. Terminal mobility is the ability to serve an identifiable terminal in different physical locations.
- *TMC.* Terminal mobility controller; it provides the functionality necessary to control terminal mobility.
- *PMD.* Personal mobility data-store; it provides database functions for personal mobility services. Personal mobility is the ability to

serve identifiable subscribers, not the terminals they are using, in different physical locations using different terminals.

- *PMC.* Personal mobility controller; it provides the functionality necessary to control personal mobility.
- *RPC.* Radio port controller.
- *RP.* Radio port.
- *RPI.* Radio port intermediary; the three radio entities (RPC, RP, and RPI) together provide the functionality of base stations and base station controllers.
- *RASC.* Radio access system controller; it is an entity providing access manager (AM) functionality (see Chap. 11 for more on AM).
- *OAM&P.* Operations, administration, maintenance, and provisioning systems.
- *IWF.* Interworking functions.
- *TE.* Terminal equipment; this is a collective representation of TE shown in the original T1P1 reference model. Since various functional pieces of TE, such as adapters, are not part of networking discussions, they are shown here as a single entity for simplicity.

Mapping of the above functional entities to the entities used in wireless networks is not complicated, except for the TMD, TMC, PMD, and PMC. These can be mapped to at least two, and maybe more, network elements. The TMC/TMD functions are accomplished in a VLR and HLR. The PMD/PMC functions are currently specified and could also be accomplished in a VLR/HLR or in some new network element. This will be dealt with by the industry when development of personal mobility applications begins.

The stage 3 (protocol) standards for the interfaces between the RPC and RASC, RPC and RPC, RASC and TMC are being developed by T1S1 under the name of the mobility management application protocol (MMAP). The protocol for the interface between the RPC and PSC is developed in TR 46. The interface standard to be finished by TR 46 also incorporates the MMAP protocol. The protocol between the PSC and various network databases is generally expected to be IS-41.

All air interface work is the responsibility of T1P1. However, since TR 46 has exactly the same responsibility within TIA, there was a potential problem of standards developed for the same radio interfaces; which would have been a disaster. To avoid it, T1P1 and TR 46 agreed to create a joint technical committee (JTC) that would develop a single set of air interface standards. This has been a very important step forward. But the variety of air interfaces that appeared in the JTC was much more than common sense would allow since 17 proposals were introduced by a number of different companies. After some deliberations, seven potential air interface alternatives were created, specifically:

- Omnipoint
- CDMA
- TDMA
- GSM
- PACS
- Wideband CDMA
- DECT

Table 4.3 presents the air interface standards developed by JTC and their official names. An air interface for each agreed technology is developed by a specific technology ad-hoc group (TAG). Some of these have already acquired ANSI standard numbers, while others are still in the works. The most progress has been made by TAGs 2, 3, 4, and 5.

DECT work is dormant in JTC, while TR 41.6 of TIA is actively involved with this technology for unlicensed applications.

Table 4.3 JTC Standards

TAG	Technology	ANSI number
1	Omnipoint (PCS-2000)*	J-STD-017
2	CDMA (IS-95 based)	J-STD-008
3	PACS	J-STD-014
4	TDMA (IS-54/IS-136 based)	J-STD-009/10/11
5	PCS-1900† (GSM-based)	J-STD-007
6	DECT	not available
7	W-CDMA	J-STD-15

* This interface is also referred to as Composite CDMA/TDMA (CCT).

† PCS-1900 is the air interface standard based on GSM, while the rest of the GSM-based standards (e.g., MAP) is called DCS-1900. Confusing? Yes, but this is the state of affairs at this time.

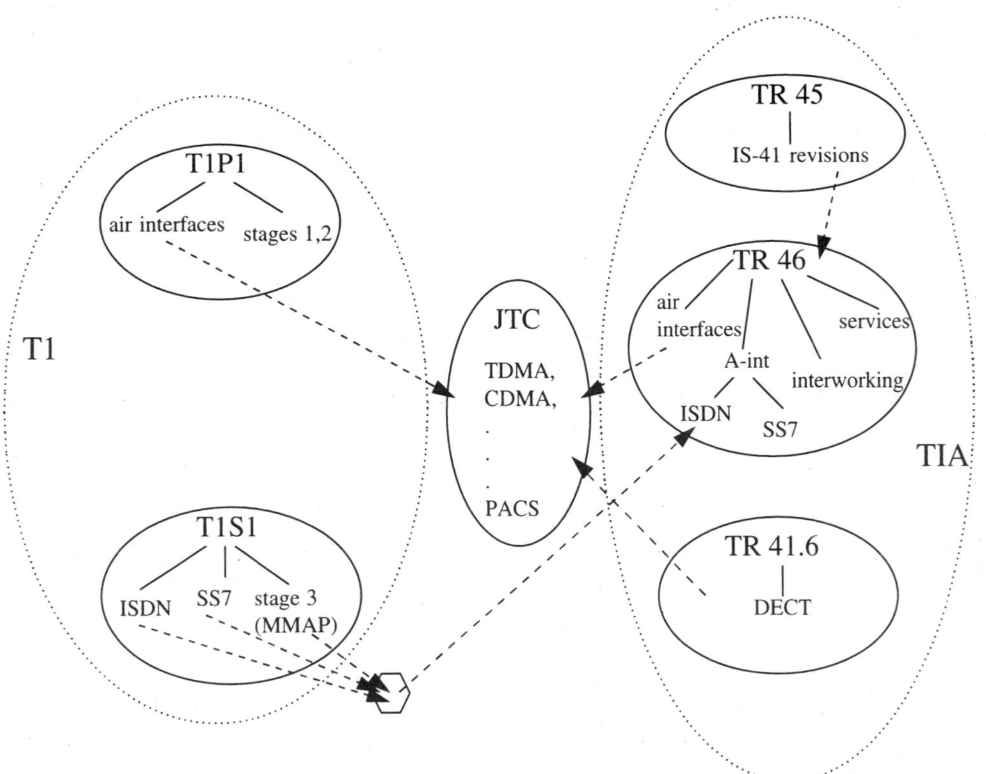

Figure 4.9 Summary of the current PCS standards activities.

Figure 4.9 shows all basic PCS standards activities as they were in the second half of 1995. And this figure does not even include the WIN standards work.

4.4 Wireless Intelligent Network

The WIN concept is basically the same as IN. We can discuss possible differences in services that each is targeting, but distributed intelligence, service creation, and triggering are the same. WIN is really a sign of the times for the wireless industry. Its development has been influenced by the surge in the number of cellular subscribers, the upcoming competition between cellular and PCS networks, and the need to offer more services to users. On the other hand, wireless networks were always operating using IN principles. What is an HLR if not remote intelligence that aids in completing calls?

Although WIN started within the cellular community, its development will also benefit the PCS community. Since cellular and PCS network architectures are basically the same and both use the same intersystem protocol, WIN is applicable to both equally.

4.4.1 Services

WIN services relate to currently existing types of wireless services as Advanced Intelligent Network (AIN) services relate to existing wireline services. In general WIN services provide more flexible and sophisticated services, offer efficiency in some services compared to those implemented without WIN, and allow a greater level of customization. The same services discussed in Sec. 3.3.2 on AIN services are projected to be covered by WIN.

4.4.2 Architecture

In our investigation of various concepts, we used open standards and existing networks to discuss their details. WIN is a little different. There is no formal standard architecture, and there are no WIN-based networks today if WIN is understood as a new network architecture which supports sophisticated services similar, but not limited, to those of AIN. However, the network architecture of cellular and PCS networks is known and is the foundation for WIN. Introduction of new network elements will depend on business and service directions for a particular network operator and will be discussed in Chap. 9. We can draw a generic architecture for WIN that will be the basis for a specific network architecture with WIN capabilities (see Fig. 4.10). This is the same network architecture used in cellular and PCS networks with the addition of a service control point and a service node

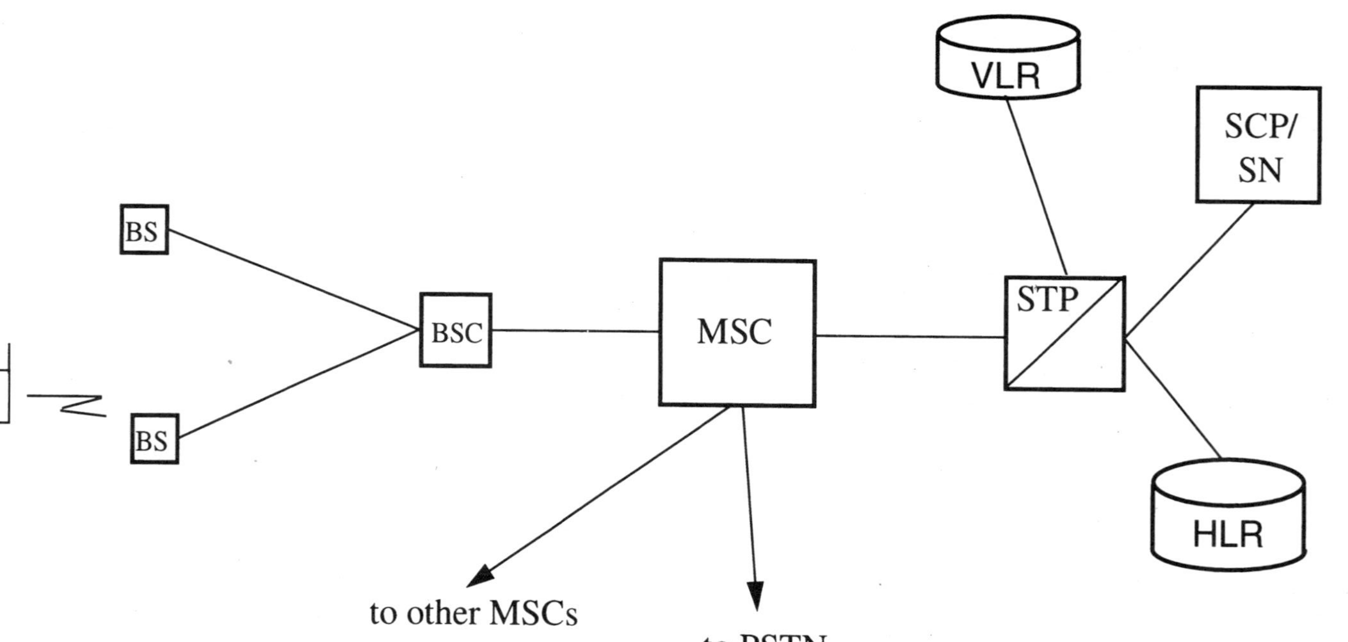

Figure 4.10 Generic WIN architecture.

(SCP/SN). Since the concept of SN is not formalized in standards or implementations, we can speak of it as a general platform for a variety of applications. As with AIN, the SCP and SN relationship will have to be clarified when networks start using these elements.

All the network elements in Fig. 4.10 are interconnected via SS7 (presence of the STP signifies this). In actual implementations the connectivity of a VLR and SN with an MSC can be of different types and depends on many factors, such as existing equipment, SN functionality, and others.

4.4.3 Standards

The work on WIN standards actually started not in a standards committee, but in the Cellular Telecommunications Industry Association (CTIA), which is an association of cellular networks providers. At the end of 1994, CTIA produced a service requirements document (SRD) which contained some modeling and some requirements for the wireless intelligent networks. This SRD was introduced in TR 45 for formal standardization. Since that time, TR 45 created a group that is responsible for the development of WIN and plans to release its first results in the form of interim standards sometime in 1996.

The call model included in the SRD was the functional foundation of the standardization work. It is a description of call states for wireless network call processing. Although basic call processing functions are the same in wireless and wireline networks, the need to handle mobility management functions in wireless networks makes it impossible to have a call model for wireless networks that completely copies the one from the wireline networks. Therefore, the initial WIN call model is different from the AIN call model.

4.5 Unlicensed Services

Twenty MHz of the "PCS spectrum" at 2 GHz are allocated for unlicensed use. The typical applications in this band are cordless phone and wireless private branch exchange (PBX). A wireless PBX is similar in functionality to a regular (wireline) PBX. This technology addresses the needs of businesses that have a large number of employees in a building. With a wireless system, these employees can move from office to office and from floor to floor and still be reached by using wireless handsets which communicate with a centrally located PBX.

The unlicensed spectrum offers great opportunities to the makers of wireless PBXs. The question of technologies capable of satisfying the needs in this spectrum is also open. Several candidate technologies exist. They are all familiar to us: PACS, DECT, and CDMA. A good candidate technology would be capable of supporting the necessary PBX-based services, effectively use low power, and comply with the etiquette in the unlicensed spectrum.

All low-power technologies are good candidates, including Wireless Access Communication System (WACS), DECT, and Personal Handy Phone (PHP)[5]—all TDMA systems. WACS is the Bellcore-designed low-power system which was a prototype for the standard PACS system. PHP is another prototype for PACS. It is a Japanese standard. PACS is a composite of WACS and PHP for the 2-GHz licensed PCS spectrum. PACS needs some modifications to work in the unlicensed band. These modifications are part of the offerings by WACS and PHP supporters but are not part of the PACS standard.

Standardization of the communications in the unlicensed band has actually started in TIA TR 41. The basis for TR 41 work is DECT, which has been renamed North American Wireless Customer Premises Equipment (NA/WCPE).

Proponents of other radio systems—IS-95 CDMA, PCS-1900, PCS-2000—also claim that their technologies can be used in the unlicensed band, and they certainly can be. However, first they have to be adopted for the unlicensed band, and, second, it is not obvious that systems optimized for use as high-mobility high-power systems can perform as well in the low-mobility low-power applications.

4.6 Multitier Concept

Cellular networks are operating using the 25-MHz bandwidth. This spectrum is used for mobile customers—subscribers who need telecommunications while they are away from their fixed telephones—when they are in a moving vehicle most of the time. All cellular services are developed to fit the 25-MHz bandwidth. With PCS the situation changes. Two types of licenses are awarded: 30 and 10 MHz. Since different bandwidth has different capacity for serving users, the question of 10-MHz usage has arisen. While it is clear that holders of 30-MHz PCS licenses can directly compete with cellular networks for mobile users, the usage of 10 MHz is not so clear. There are several alternatives:

1. If a 10-MHz license is acquired by a current cellular provider, this bandwidth can be added to the existing 25 MHz for added capacity.

---

[5] It is sometimes referred to as Personal Handy Phone System (PHS).

2. This bandwidth can be used to compete with cellular providers for mobile users.

3. This bandwidth can be used to serve a different set of customers, those who are not moving fast (in cars) but are either stationary or moving slowly, maybe at pedestrian speed.

Since there is a lot of interest in pursuing the third alternative, this kind of service received a name that distinguishes it from the regular mobile market segment. This is now called a low-mobility service, whereas a regular mobile service is called a high-mobility service. Although there are differences in the amount of spectrum and power required for low and high mobility, the main distinguishing characteristic is the speed of movement. In low mobility the speed is low (pedestrian or, possibly, very low vehicular). In high mobility the speed is high while a vehicle is moving. High-mobility service can definitely serve slow-moving users, when they happen to move at a slow speed or not at all. But low-mobility service can do it cheaper, and, maybe, better, by specifically addressing a low-speed market.

We are not trying to prove the economics of different types of services, but only to define a PCS concept, which is called a multitier service. Let's suppose that a subscriber uses the phone in a car at high speeds and pays appropriate fees for the service. When the subscriber's vehicle is stuck in traffic, let's say for an hour, the subscriber switches to a cheaper service which can only be used at low speed or while stationary. This subscriber uses services of both high- and low-mobility providers to gain an economic benefit. This is basically a definition of a multitier service: a service that switches subscribers between a high- (high tier) and low-mobility service (low tier) as required.

As we discussed before, wireless networks were originally developed to serve subscribers moving at vehicular speed. The radio and network technology was geared to that particular type of service. Some may say that radio communication is radio communication and the system designed for handling mobile subscribers can easily deal with slow-moving subscribers. Proponents of all available radio technologies will say that their systems, GSM, CDMA, etc., can handle what we call low-tier service. While it is true that these systems are capable of supporting low-tier services, they probably cannot do it in a way that would satisfy the need of this new market segment. Some of these needs can be identified as low price, light battery, and voice quality comparable to wireline service. To satisfy these needs, specific technological solutions should be found; otherwise the market may not bear the costs or other inconveniences and may never generate significant

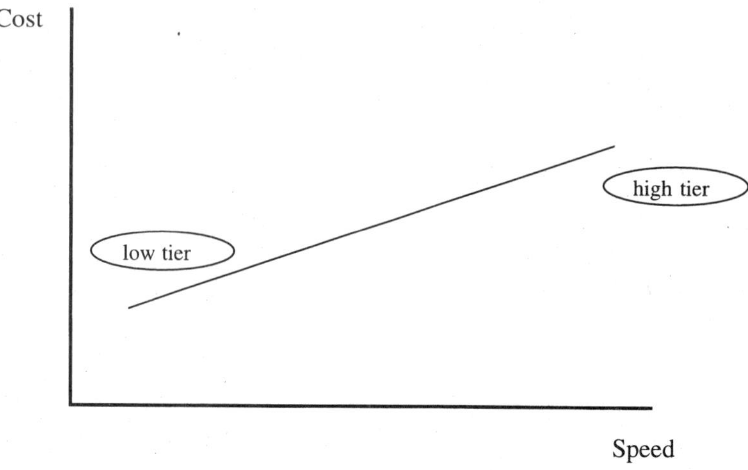

Figure 4.11 Relationship of speed and cost of service.

demand. The graph in Fig. 4.11 shows the relationship between speed and cost of service.

Some of the radio technologies that we mentioned before have actually been created to support low-tier services. These are PACS and DECT. There are also other candidates, such as PHP and CT-2. The other radio technologies should be considered for primary applications in the high tier.

If two types of technologies exist to support two environments, it is reasonable to question whether a combination of two services is possible. If it is, as we discussed in the example above, we can speak of the concept of a multitier service.

Chapter

5

Why Seamless Networks?

Everybody is talking about convergence—convergence of computing and telecommunications, convergence of wireline and wireless networks, convergence of entertainment programming and its delivery, convergence of national and international communications. The world that used to consist of fragmented industries, such as computers, entertainment programming, cable TV, telecommunications, and others, is now bracing for a big change. Only recently (mid-1995), Disney and Capital City/ABC, and Westinghouse and CBS merged, and there were other successful and unsuccessful merger attempts between telecommunications and cable TV companies. By the time this book is published, there will be more of this kind of news to deal with.

A new telecommunications law (Telecommunications Act of 1996) has been adopted in the United States. Its adoption follows a long debate among lawmakers concerning existing laws and regulations, which came about because of the readiness of the industry and the public to accept a change in existing law. Although not every stakeholder is completely satisfied with the implications of the new law, it is clear that it will encourage more competition and convergence.

A process that will result in a more competitive environment is underway in Europe. Major European countries, such as Germany and Italy, are privatizing their telecommunications to allow more competition. The European Community agreements actually require this new openness. This and the opening of new big markets, such as Russia, in other parts of the world provide a foundation for the future development of worldwide services which would create opportunities for global service companies to become successful. The result for some American companies may be to move toward an international footprint.

The issue of integrating wireline and wireless networks, making them seamless, is just one aspect of the overall convergence thrust. However, there are some specific trends in the telecommunications industry that are driving wireline and wireless networks toward each other. The results of these trends are business needs that we will discuss. Although it is hard to make a somewhat artificial division of needs, it is useful to group them into competitive, service, and technical categories. The point here is not to come up with this classification, but to use it as a tool to understand the business needs at different stages of the decision-making process. We will use this particular paradigm to approach the issue of seamless networking: first, a competitive need is established, then service(s) that are important in a particular business environment are defined, and after that technical implications are analyzed.

5.1 Competitive Needs

The competitive issues of seamless networking are related to the growing and varied wireless market. Companies that are in the wireless business compete for customers, and therefore, coverage areas. A so-called national footprint is often mentioned as a goal of the existing or new wireless companies. A national footprint is an ability to serve customers nationwide. A company that has a national footprint is in a better competitive position than otherwise. The question is how to achieve this. One way is to acquire operating licenses across the country in places sufficient to cover customers anywhere in the United States. While this is theoretically possible, the amount of money involved and other business issues may make such an approach difficult. Another possibility is to create alliances with other service providers, wireless and wireline, and together achieve a goal of a national footprint. A third possibility is to build some limited infrastructure and lease equipment, in a form of services, from other sources, such as third-party service providers, friendly complementary companies, or even competitors who would sell spare capacity available on their systems.

In any of the above cases there may be a need for arrangements to make various involved networks seamless. A company growing a national presence out of separate networks it owns, may have different equipment using dissimilar protocols. This would have to be changed if the goal of this company is to provide feature transparency nationwide. In addition, this company will have to interwork with public network(s) for routing, call delivery, and possibly other services. So, seamless networking would also be helpful. In the case of several companies working together to establish a national footprint, the need to create a seamless network is even more evident since there will be differences among those networks.

The above discussion pertains to service aspects of competition and derives the need for seamless networking from the requirement to establish a better, more competitive, service model. Another aspect of competitiveness is operating costs: the lower operating costs are, the easier it is for a company to compete. Sometimes, cost reduction can be achieved through the use of somebody else's equipment instead of investing capital and buying this equipment. This is called *leasing* in some businesses. In the world of networks, this would be *subscribing* to someone's network services, such as services that require specialized databases. For example, company A needs databases, but, instead of buying them, uses database services from company B. When such an arrangement is negotiated, the need to also negotiate technical interworking issues is obvious. A network which is seamless between companies A and B would make things a lot simpler.

If a company wants to establish an international presence and create a footprint covering more than the domestic market, it must interwork with networks, either old or new, which are built using different standards. This requires an understanding of the technical issues and then solving them. These are issues discussed further in this chapter. Just imagine a U.S. wireless company, which operates a code division multiple access (CDMA)-based network, that wishes to integrate this operation with a Global System Mobile (GSM) network somewhere in Asia. The incompatibilities between the two networks would have to be resolved if the resulting network is to operate in a seamless fashion.

5.2 Service Needs

In general, services drive the evolution of networks and their interworking. Any service scenario may require support by a seamless network. Let's discuss some service-oriented issues requiring seamless network support.

Feature transparency is the need to hide from a customer details of what kind of a network (technology, etc.) provides service at any time. If a customer uses both wireline and wireless services, it would be beneficial to allow the customer to use the same services in exactly the same way regardless of the network the customer is served by at the moment (wireline or wireless).

Roaming is a critical service because it allows users to use wireless phones anywhere in the country. In addition, feature transparency becomes very important as a roamer wanders across different networks with different technologies but would like to use services (whatever are subscribed to) the same way as at home.

Number portability as a service would require a number to follow a user. The number has to be recognized by different networks, and a seamless network would make things a lot easier.

Another service-oriented issue is service efficiency. It may be needed, in order to provide a service efficiently, to cross-use databases located in adjacent networks. Access to these databases would be simplified if seamlessness is achieved.

Ultimately, a service goal of any network operator thinking globally is to provide universal services. Such services are being defined by international standards committees and will be very attractive to users, given an affordable price and service convenience. A seamless network is a must to make these services work.

5.3 Technical Needs

When two different networks interwork, many technical issues arise. The type of interworking is based on the business environment that forces the interworking, but these types can be categorized into one of two groups of issues: interface/protocol issues and system issues.

The interface/protocol category consists of issues of harmonization of the language (one of the specific protocols used at a specific layer) used between two interworking networks. If two networks use even slightly different variations of the same language (protocol), no understanding can be achieved without extra efforts.

The system category consists of issues which are based on the differences that exist between similar systems deployed by different networks. For example, if databases in two networks contain the same type of information (e.g., service profile of users), but the database in one network generates a specific command to another network element (e.g., a switch) when it is queried, while the database in another network does not do that, a user will see some difference in the service (timing, a possible announcement, etc.). Such a result is not good for either a user or a service provider.

The difference between the two categories can be summarized as follows: The interface issues are of the communicating type and the system issues are of the application type (see Fig. 5.1). Examples of systems issues are behavior of switches and databases, and examples of interface issues are signaling protocols. A specific interworking arrangement may have to deal with either or both types of issues.

5.4 Issues

Seamless networks is a very nice goal, but many issues have to be studied and solutions to many discrepancies will have to be found to

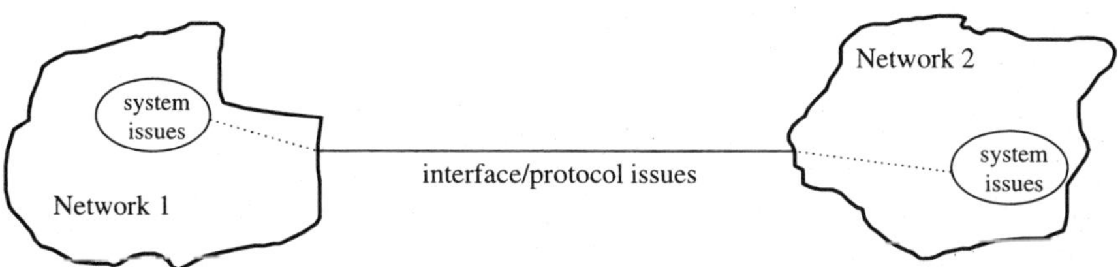

Figure 5.1 Categories of technical issues.

achieve this goal. Some solutions will definitely require cooperation by many interested parties; others may be local, applied only in a specific arrangement by a service provider in need of a specific or quick solution. It would be hard to predict right now many of these solutions, but it is possible to start listing the issues and discussing their implications. Whether one or another solution is found in a specific case is not as important to us now as understanding what seamlessness means and where the potential pitfalls and opportunities are.

Making seamless networks is a very complex task. Figure 5.2 illustrates a variety of different components of a potential seamless network that have to be taken into account. Their variety clearly points to the need to understand different aspects of all these components, which can only be achieved through careful comparison and analysis. The first step in this process is identification of the different issues to be studied.

The investigation of system and interface issues affecting network interoperability would be aided if some form of classification is devised. Although many ways for such classification exist, the goal is to simplify the problem by addressing one category of issues at a time and trying to understand the relationship between different categories. One classification is as follows:

- Intelligent network issues focusing on intelligent network (IN) capabilities, including service creation, protocols, and architectures
- Interconnection issues, including various aspects of Signaling System #7 (SS7)
- Operations (network management) issues
- Network access issues, specifically addressing various techniques aiding in the convergence process
- Global System Mobile (GSM)-related issues
- Service transparency
- Standards as a prerequisite for network integration

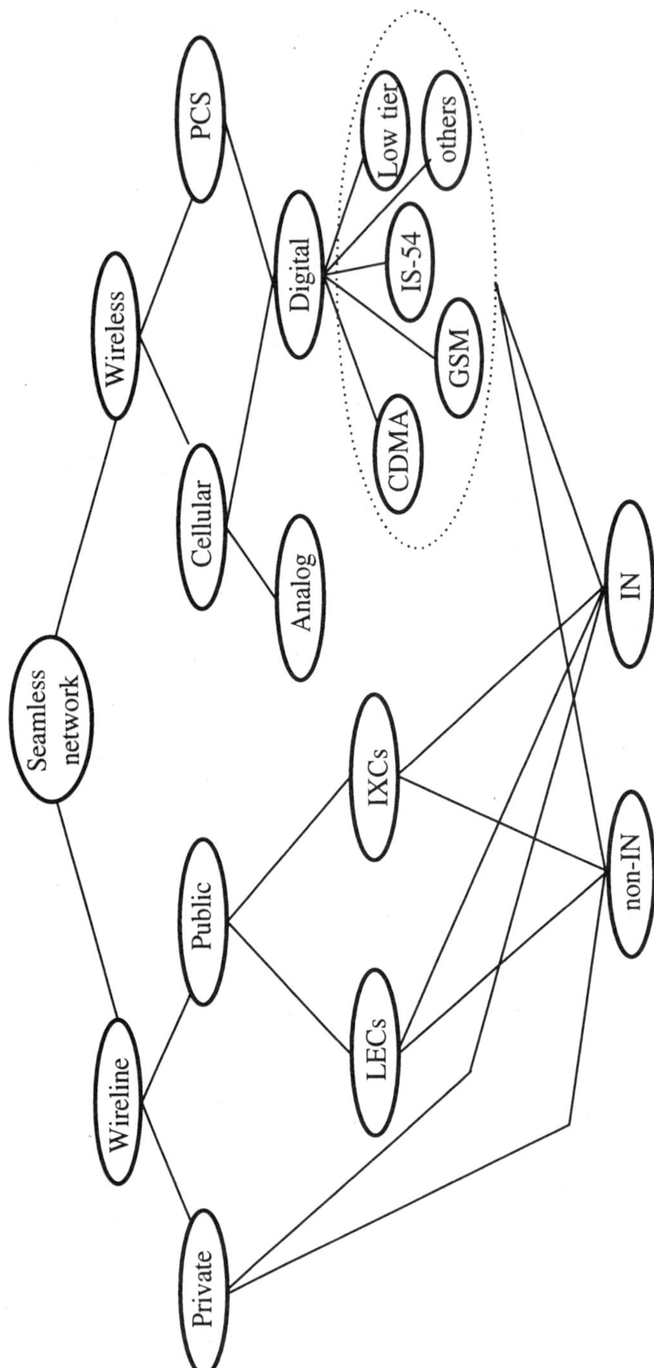

Figure 5.2 Complexity in creating seamless networks.

The following chapters present different approaches to making networks seamless according to the above classification. Some of these approaches cover more significant aspects of seamlessness than others, but every approach adds to the process of breaking barriers among differently built networks. A combination of different approaches is possible. Their selection depends on local conditions, such as the business environment, market pressures, type of competition, installed equipment base, etc. Each integration approach involves a recognition of issues that require solutions in order to achieve success. We will discuss the issues that surfaced so far.

Chapter

6

Intelligent Network Issues

An intelligent network (IN) is the key to providing sophisticated services in both wireline and wireless networks. The components of IN are switches, peripherals, and databases that provide capabilities based on which an innovative service can be defined. Moreover, these IN components are the platforms that can be used for a great number of new services as they are being conceived, developed, and introduced by service providers. Since both wireline and wireless environments are clearly moving toward IN, integration of functions between Advanced Intelligent Networks (AINs) and wireless intelligent networks (WINs) is critical in the development of a seamless network. This is important for a couple of general reasons. First, architectural integration would be driven by IN components, primarily databases [service control point (SCP), home location register (HLR)] and service nodes, including intelligent peripherals. Second, it is easy to imagine that support for seamless services would require a wireless network to query wireline network's SCPs and vice versa. Databases provided by a third party may also play an important role in network operations, and, therefore, accessing multiple IN-type remote databases may be needed from a single network.

The following are IN-related issues that will affect seamless operations:

- *Modeling issues.* These affect current and future services; modeling is the foundation of practical implementations, such as protocols.
- *Applications issues.* These are primarily protocol and service creation issues, and they also affect services, but from a practical point of view; something can be specified in a model but not be supported by a protocol or service creation techniques.

- *Interconnection issues.* These are related to Signaling System #7 (SS7) since this Common Channel Signaling (CCS) protocol is the main carrier of IN messages
- *Network architecture issues.*

All of the above types of issues affect the network architecture that would support seamlessness. The architectural alternatives that can be identified and used to advance seamlessness relate to the main components of IN—switches and databases.

In this chapter we will look at the modeling differences between an intelligent network that is being used in the wireline world (AIN) and the one which is planned for the wireless world (WIN). These differences affect call processing and can be visible in the call models, specifically, in the types of triggers that are defined. One difficulty in performing this comparison stems from the tentative nature of the WIN document which is available at the time of this writing. It is possible that by the time this book is published WIN models and protocol will have been changed and will vary from the ones discussed here. However, it is safe to assume that the development of WIN standards will be based on the currently available models. Our analysis, even if it is not precise given the tentative nature of documentation at hand, is still valuable since it provides intermediate results and tests the methodology for such comparison. It is also valuable to show how differences between AIN and WIN could affect seamlessness.

6.1 Call Models

A *call model* is an abstraction of sequential functions that play a role in call processing. Development of a call model advances the understanding of the call processing in a given environment, at the same time formalizing it. This makes design of an extremely complicated call processing software easier to manage. A call model also aids in the development of additional functions to support new services.

Although basic call processing seems to be simple to comprehend, call models that describe call processing can be of different types. A simple call processing algorithm can be as follows:

> call attempt identified => digits collected => outgoing trunk found => call connected => disconnect encountered => call cleared

This is a simplification, but it demonstrates the point that a call model is an algorithm and its definition affects the operations of call processing. The kind of a call model used in the network makes an even bigger difference when more complex services are developed. If a call model treats each global (end-to-end) call as an object, some centrally located system has to be aware of all components (legs, connections) of a call. For example, call processing in many private networks using private branch exchanges (PBXs) is modeled this way. Call processing in public networks is usually modeled using a different approach, which is referred to as a *single-ended view*.

A single-ended view treats calls as half-calls: one-half is originating and another half is terminating. The reason for this lies in the complexity of public networks. If a network were simple, let's say consisting of a single switch, the call model would definitely be of a global-view type. In this model a switch would know everything about origination and termination for a particular call, as in Fig. 6.1.

The real networking situation is different in most cases and is shown in Fig. 6.2. Usually the originating switch is aware only of information pertaining to call origination, unless a sophisticated signaling system is available to carry all information about call termination from the terminating switch to the originating switch. And, if the originating switch is completely in charge of a particular call, all additional services at the termination side would have to be handled by the originating switch. These are some of the reasons why the call model in the public networks is a single-ended-view type. The result of this is call processing separated into originating and terminating, each handling only one portion of a call.

The scope of coverage that would have to be required with the global view of a call in a more complex network arrangement is shown in Fig. 6.2 as a possible global view.

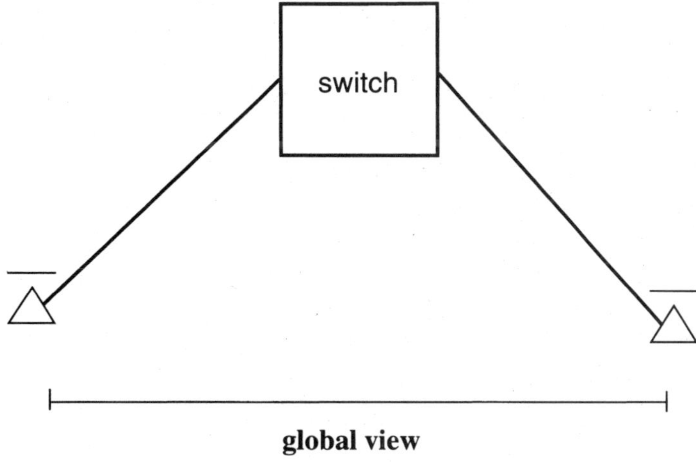

Figure 6.1 Global view of a call.

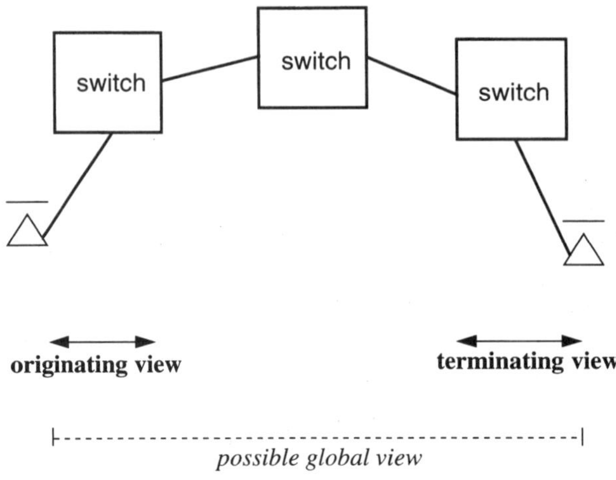

Figure 6.2 Single-ended view of a call.

6.1.1 AIN call model

The AIN call model has been developed based on the call processing needs of a wireline network and is of a single-ended-view type. This call model shows call states (they are formally called *points in call*—PICs—but we will refer to them as states) and certain detection points, which are part of the triggering mechanism discussed in Chap. 3 (see Figs. 6.3 and 6.4). Each call state has a name that signifies its purpose. Smaller boxes with numbers inside are detection points for the triggering.[1] The numbers are as defined in the AIN requirements.

Two types of detection points are defined in the AIN call model: trigger detection points (TDPs) and event detection points (EDPs). Both specify places in the call model where certain events can happen through the activation of specified triggers. The difference between them is in the provisioning. TDPs are provisioned with *triggers* in the service switching points, or SSPs (switches). EDPs are provisioned with events from the SCP at the time that an application requests that such an action be performed in the SSP. Usually events are sent from the SCP to the SSP in the response to the request message that was originated by a trigger (see Fig. 6.5).

EDPs can also be armed by an SCP through the *Create_Call* message. This message can be issued by an application in the SCP, and some of the parameters carried by this message are the events that the application is interested in.

---

[1] Shaded detection points in the AIN call models are those defined for later AIN releases and generally are not yet available in products.

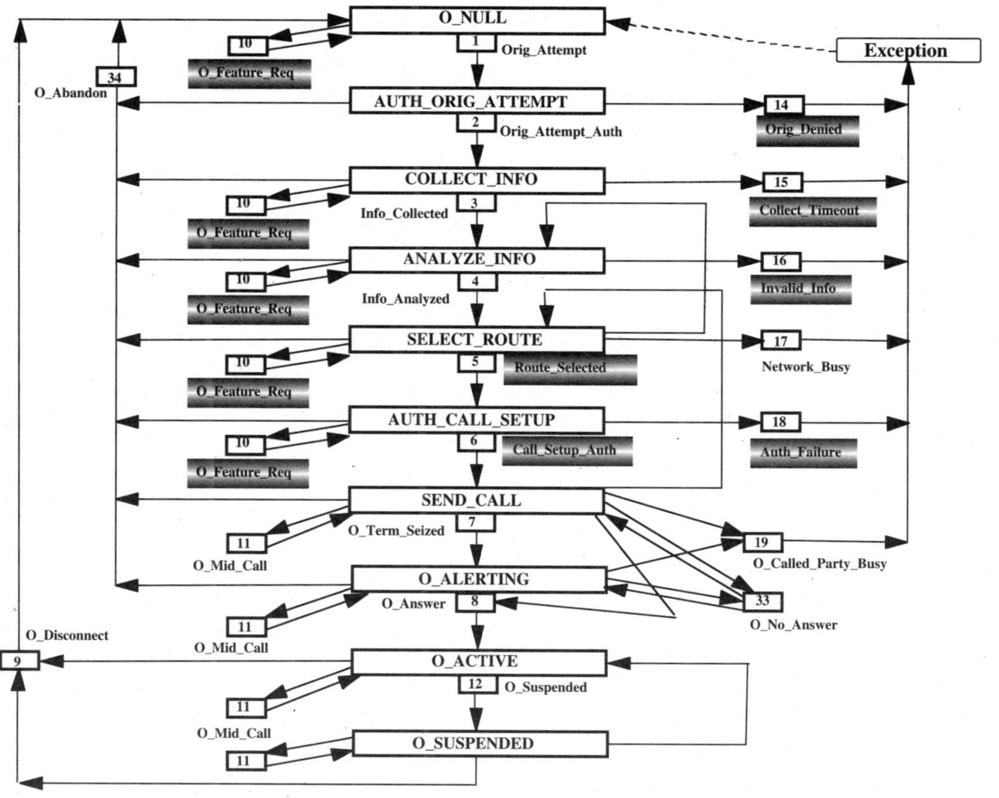

11 - TDP/EDP; the number indicates an AIN number assigned to this particular detection point.

Figure 6.3 Originating AIN call model.

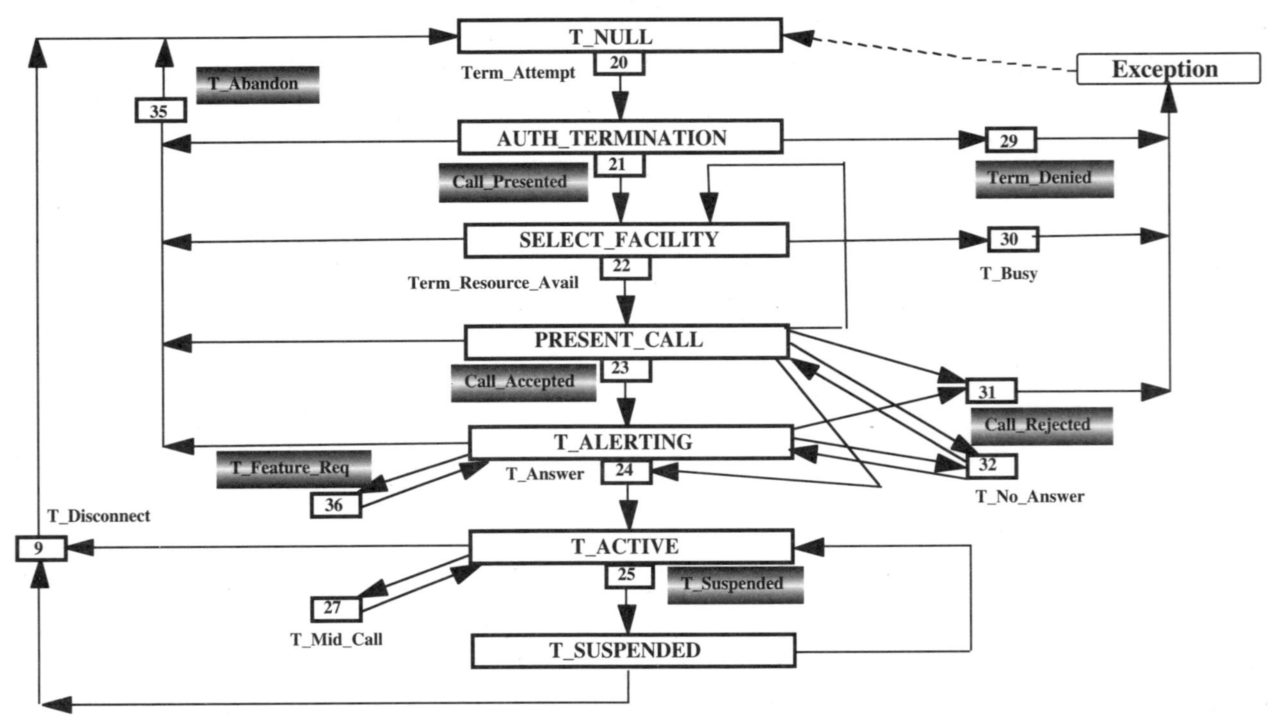

Figure 6.4 Terminating AIN call model.

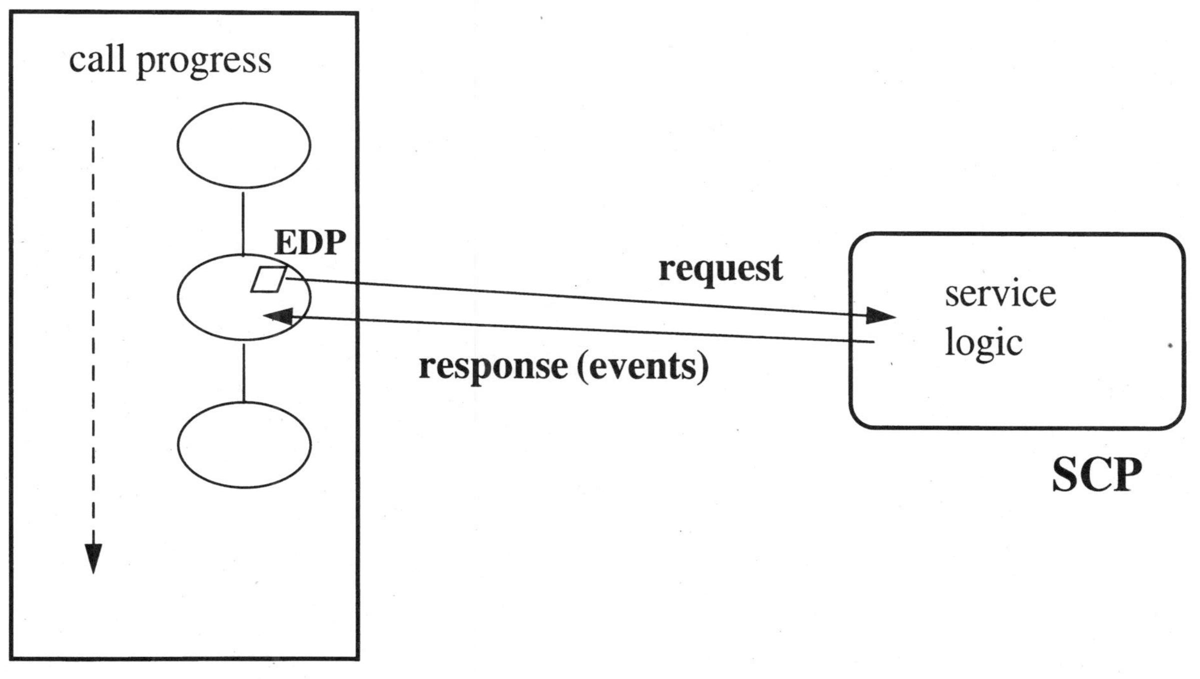

Figure 6.5 Installing event requests and notifications.

Events at EDPs can be of two types: requests and notifications. Requests work the same way as triggers; that is, call processing is suspended when a request is encountered. The difference, as we said, is in the provisioning of them. Notifications are different. When encountered, call processing is not suspended. Instead, reports about occurring events, specified in the EDP, are issued to the application in the SCP. This notification process is used to allow an application to monitor activities of interest at the switch and make certain adjustments or use the reports to accumulate statistics needed by an application collecting them.

For example, an application may need to accumulate statistics about specific types of originating calls [originating directory number (DN) or destination country or something else]. With a notification event installed, call processing would not be suspended when a message with the report is issued to an SCP.

6.1.2 WIN call model

The WIN call model has been developed by the Cellular Telecommunications Industry Association (CTIA) and may be modified by TIA TR 45. It generally follows the call model developed for public networks. Mobility requirements put additional demands on call processing and the call model for the wireless environment, and they make the two call models not exactly the same, but compatible. Figures 6.6, 6.7, and 6.8 show the WIN mobility management, originating and terminating call models.[2]

A comparison of the states of AIN and WIN models shows a lot of similarities. Both use the same type of call processing from information collection to analysis, mid-call interventions, and disconnect. The most obvious differences are in the mobility needs of WIN, which are represented by an additional model for mobility management. It shows states while no call processing is occurring, that is, while authentication and registration are taking place outside of call origination or termination. AIN does not need such a model because these kinds of activities are not occurring in wireline networks (not yet, anyway). There are also a couple of other differences.

The first difference is derived from the nature of the wireless network activities. These activities are inherently IN-like. While the distribution of intelligence is a new requirement for wireline networks, in the wireless networks it is a way of life. The unstable nature of subscribers' locations requires centralized intelligence to keep track of

---

[2] Shaded detection points in the WIN call models are those which, according to the WIN service requirement document (SRD), are implicitly included in IS-41.

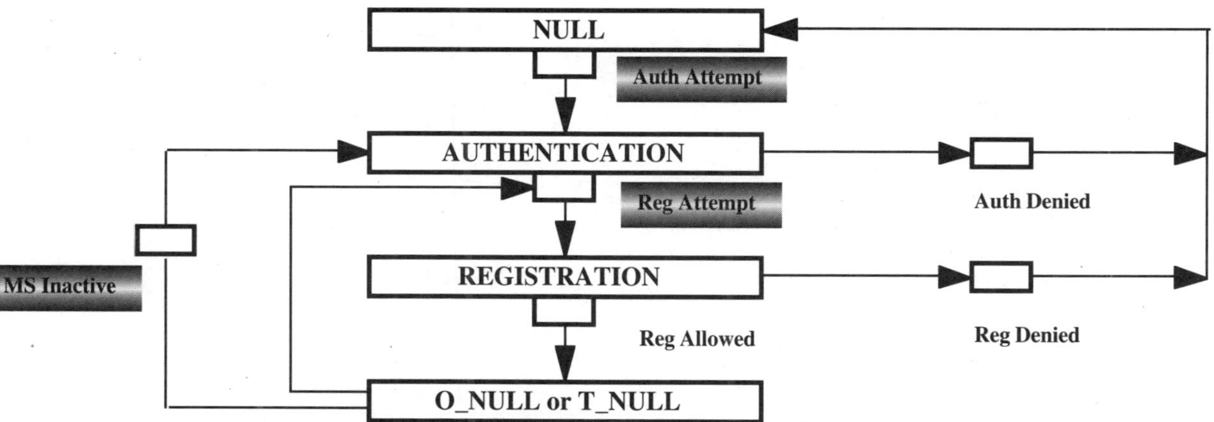

Figure 6.6 Mobility management model.

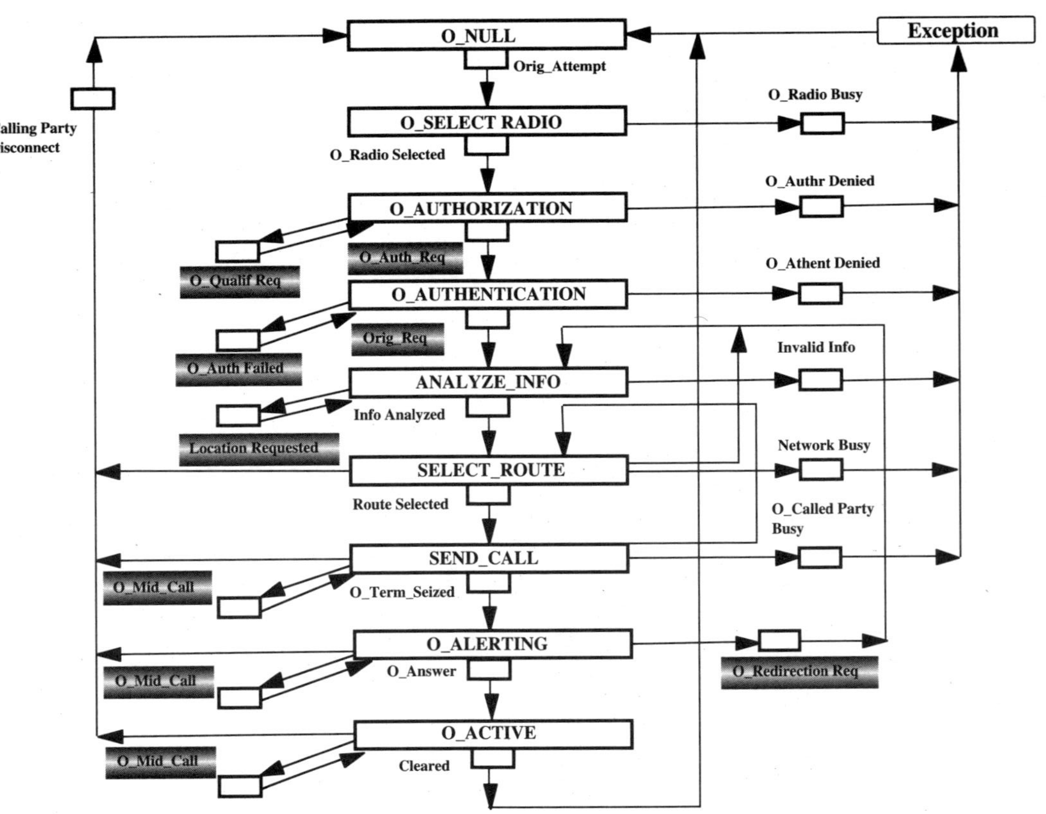

Figure 6.7 WIN originating call model.

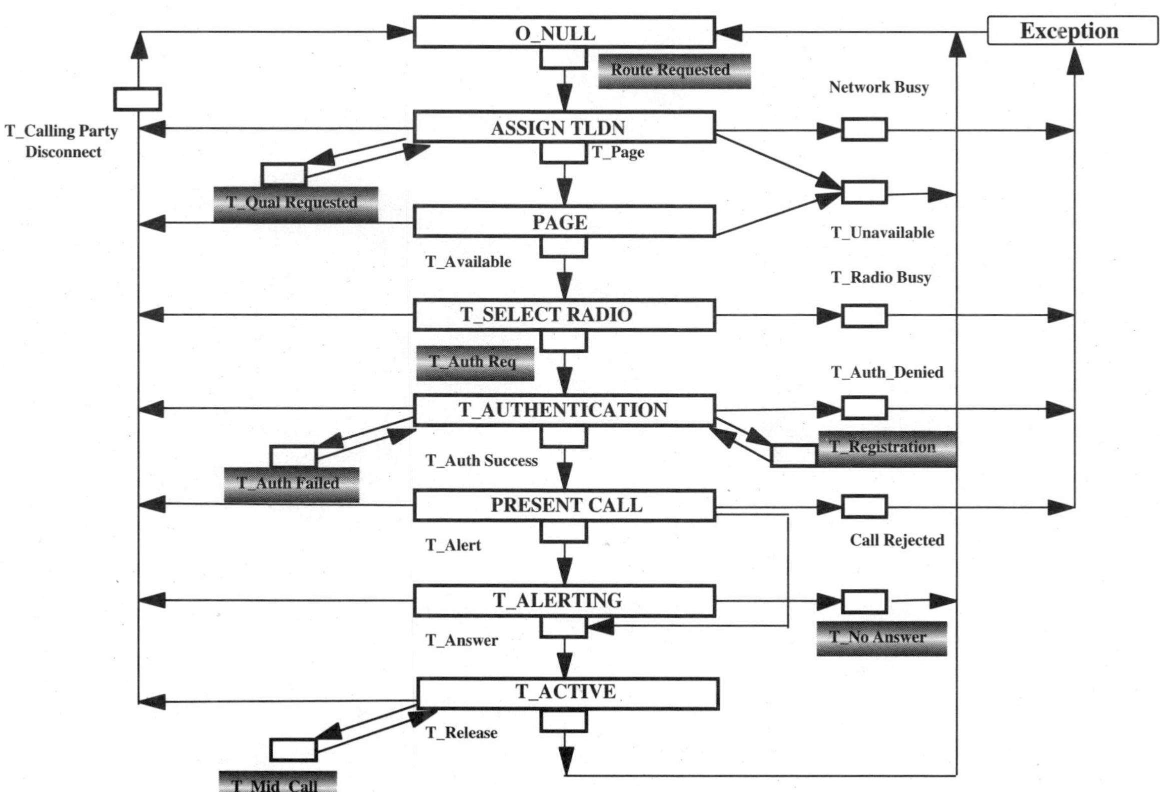

Figure 6.8 WIN terminating call model.

them. Wireless networks behaved according to some intelligent network principles before the WIN work started. When it did start, many of the TDPs shown in the WIN models were just documented without imposing additional requirements on the mobile switching centers (MSCs). These are all shaded TDPs in the models.

There are also some differences in the call states (besides their names) of AIN and WIN models. For example, the *O_Select Radio* state in the WIN originating call model does not have a counterpart in the AIN call model because in the wireline network there is no need to find a communication channel to a subscriber. Another example is the *O_Suspended* state in the AIN call model. A call appears in this state after it was in the *O_Active* state and a terminating side was disconnected. The call is being suspended until the calling party disconnects or the terminating party reconnects. This state does not exist in the WIN call model. While the call is in the *O_Active* state (WIN), either a calling or called party disconnect results in a transition to the *Null* state. So, the AIN call processing allows a terminating party reconnection, while WIN does not.

In general, though, we can state that the discrepancy between the two call models is manageable, especially since the WIN work is still going on. Comparison of available or planned triggers would tell us more about potential service-related differences.

6.2 Triggering

Since the AIN and WIN call models cannot be exactly the same, but are sufficiently alike, it is important to investigate the type of TDPs and triggers that each model provides. By analyzing the triggers available in each environment, a conclusion about the type of services that the environment supports can be reached. As will be seen from the list of triggers in this chapter, they are not totally similar. Different placement of the same triggers can also result in different implementations of the same services and, therefore, can cause a problem for a seamless service.

6.2.1 AIN triggers and events

Detection points are listed in two separate categories: originating and terminating. In most cases a detection point corresponds to a trigger (or an event). In some cases multiple triggers are assigned to a TDP. An encountered trigger or event results in a message being issued to the SCP with information describing the occurred activity. A detection point, as specified in the AIN call model, can be a TDP, an EDP, or both. For the purposes of the list that follows, when a detection point is an EDP, it is a request event unless a notification is explicitly stated. This list does not include the shaded TDPs shown in the call models because they are identified for future AIN releases. Sufficient information for their comparison with the WIN triggers is not currently available.

Originating call

Origination attempt. Detected when off-hook action has been performed (TDP/EDP).

Origination_Attempt_Authorized. Detected when an authorization for placing calls has been verified (TDP).

Info-Collected. Detected when enough information is collected to process the call (TDP). Three triggers are specified, and their usage is tied to the interface over which the information is being collected:

- Off-hook_delay trigger is encountered on non-ISDN lines, ISDN lines, or private facilities.
- PRI_B-Channel trigger is encountered on a dedicated B-channel on primary rate interface (PRI).
- Shared-Interoffice_Trunk trigger is encountered on public trunks.

Info_Analyzed. Detected when specific feature indicators are received (TDP). These indicators are codes or strings of digits. Seven triggers, varied by the type of information being analyzed, are specified:

- BRI_Feature_Activation_Indicator trigger is encountered when an Integrated Services Digital Network (ISDN) feature activation indicator administered at the SSP has been received.
- Public_Feature_Code trigger is encountered when a service code has been received (e.g., *57 causes the trigger to be encountered for all customers).
- Specific_Feature_Code trigger is encountered when a specific service code has been received (e.g., *57 is associated with one customer, while *62 is associated with another customer).
- Customized_Dialing_Plan trigger is encountered when an access code or a 1-to-7-digit intercom code has been dialed.
- Specific_Digit_String trigger is encountered when an appropriate digit string has been dialed.

- N11 trigger is encountered when a designated N11 (e.g., 611) code or a designated 3-digit number has been dialed.
- Network_Services trigger is encountered when a Network-Specific Facilities information element has been received in the ISDN SETUP message.

O_Term_Seized. Detected when an indication of a call acceptance has been received from the terminating side (EDP—a notification event).

O_Answer. Detected when a notification has been received from the terminating side of a called party answering the call (EDP—a notification event).

O_Disconnect. Detected when a disconnect from the calling party has been received (EDP).

O_Mid_Call. Detected when a mid-call interruption is initiated by a user at the originating switch or AIN (TDP/EDP). Three triggers are specified:

- O_Switch_Hook_Flash_Immediate trigger is encountered when a switch-hook flash indication has been received from a non-ISDN line.
- O_Feature_Activator trigger is encountered when an indication is received from an ISDN (basic rate interface, BRI), line of the designated feature activator indicator.
- O_Switch_Hook_Flash_Specified_Code trigger is encountered when a switch-hook flash indication has been received from a non-ISDN line and additional information is collected after a dial tone is issued by the switch.

O_Suspended. Detected when a notification of a called party disconnect has been received from the terminating side (EDP).

Network_Busy. Detected when trunks are unavailable (TDP/EDP).

O_Called_Party_Busy. Detected when an indication is received that a called party is inaccessible (TDP/EDP).

O_No_Answer. Detected when a called party is being alerted and the timer for this process is expired (TDP/EDP).

O_Abandon. Detected in one of the first eight states when it is set by an SCP (EDP).

Terminating call

T_Disconnect. Detected when a disconnect from the calling party has been received (EDP).

Termination_Attempt. Detected when a call appears at the terminating switch (TDP).

Term_Resource_Available. Detected when it has been determined that the terminating access is available (TDP/EDP—a notification event).

T_Answer. Detected when the called party has answered (EDP—a notification event).

T_Mid_Call. Detected when a mid-call interruption is initiated by a user at the terminating switch (TDP/EDP). Three triggers are specified:

- T_Switch_Hook_Flash_Immediate trigger is encountered when a switch-hook flash indication has been received from a non-ISDN line.
- T_Feature_Activator trigger is encountered when an indication is received from an ISDN (BRI) line of the designated feature activator indicator.
- T_Switch_Hook_Flash_Specified_Code trigger is encountered when a switch-hook flash indication has been received from a non-ISDN line and the additional information is collected after a dial tone is issued by the switch.

T_Busy. Detected when the terminating access is not available (TDP/EDP).

T_No_Answer. Detected when the called party does not answer within a specified time (TDP/EDP).

6.2.2 WIN triggers

The WIN call model contains only TDPs and triggers; there are no EDPs. Lack of EDPs and associated requests and notifications also disallows application processes that would require accumulation of statistics without suspension of call processing. The available TDPs can be grouped as follows:

- Mobility related
- Call control related

Mobility-related triggers are natural triggers that occur in the process of authentication and registration. Even before the WIN work started, the cellular networks identified certain points in the processing of mobility-related information at which an MSC queried a visitor location register/home location register (VLR/HLR) for information. TDPs and triggers documented in the WIN documentation at the time of this writing are listed here. Some of them do not yet have associated messages specified. However, while the WIN document is clearly not completed yet, the attempt here is not to be exhaustive but to identify principal directions of WIN that can be used in making comparison with AIN specifications. The list of TDPs is in three parts.

Mobility management

Authentication attempt. Detected when a mobile terminal is attempting to register. It generates a message to the HLR requesting authentication.

Registration attempt. Detected when registration is requested. A message to the HLR requesting registration is generated. Three triggers are specified:

- Registration notification trigger is encountered when the authentication is successful.
- Authorization period trigger is encountered when the authorization period for a handset has expired or when an autonomous registration occurs.
- MS active trigger is encountered when a handset that previously had an inactive status attempts to register.

Registration allowed. Detected when the registration of a handset has been completed.

MS inactive. Detected when a mobile station (MS) has been marked inactive.

Authentication denied. Detected when the authentication of the MS has not been successful.

Registration denied. Detected when the registration of a handset is denied.

Originating call

Originating attempt. Detected when a handset attempts to originate the call.

O_Radio selected. Detected when a traffic channel has been assigned to a handset for communications.

O_Qualification requested. Detected when no appropriate handset profile information is found at the MSC. A message requesting handset qualification is sent to the HLR.

O_Auth_Request. Detected when a call is being originated by a handset for which an authentication is required (e.g., it is required for each call). A message to the HLR is issued requesting authentication.

Originating Request. Detected when authentication is successful and call processing can resume. A message to the SCP is issued.

O_Auth Failed. Detected when authentication performed locally by the MSC using shared secret data (SSD) has failed. This may be reported to the HLR.

Info analyzed. Detected when all necessary information is collected by the MSC. A message is issued to the SCP. Three triggers are specified:

- Information analyzed
- Carrier identification code
- Dialed information

Location requested. Detected when the dialed digits correspond to a valid destination DN.

Route selected. Detected when the MSC has determined the origination routing information for the call.

O_Term seized. Detected when an indication of the called party alerting has been provided.

O_Answer. Detected when an indication of the answer by the called party has been received.

O_Redirection requested. Detected when an indication of the call redirection has been received from the terminating side.

Cleared. Detected when an indication of a terminating party disconnect has been received.

O_Mid call. Detected when a flash-hook or a SEND key is pressed on the handset. A message is issued to the SCP with a feature request. Two triggers are specified:

- *Feature request.* Encountered during feature activation/deactivation.
- *Hold trigger.* Encountered when the called party has been put on hold.

O_Calling party disconnect. Detected when the calling party disconnects.

O_Radio busy. Detected when no radio channel can be assigned to the originating handset.

O_Authr denied. Detected if the handset is not authorized to originate calls according to the profile at the MSC.

O_Auth denied. Detected when the authentication of the handset has not been successful.

Invalid information. Detected when collected information is not valid.

Network busy. Detected when no routes are available for call setup.

O_Called party busy. Detected when an indication has been received that the called party is busy.

Terminating call

Route requested. Detected when a request for temporary line directory number (TLDN) has been received by the terminating MSC. The TLDN is sent to the HLR.

T_Qualification requested. Detected when no profile for the terminating handset is available at the MSC. A message is issued with the qualification request.

T_Page. Detected when a TLDN has been assigned. Paging is initiated.

T_Unavailable. Detected when it is determined that the handset is not available for paging.

T_Available. Detected when a handset responded to the page.

T_Auth requested. Detected when the MSC is not capable of performing authentication of the handset. A message to the HLR is issued with the request to authenticate.

T_Registration. Detected when no registration information is available at the MSC. A message requesting the registration is issued to the HLR.

T_Auth failed. Detected when authentication of the handset by the MSC failed.

T_Auth success. Detected when authentication of the handset by the MSC is successful.

T_Alert. Detected when an indication of call acceptance has been provided.

T_Answer. Detected when a handset has answered.

T_Release. Detected when the called party disconnects.

T_Mid call. Detected when the called party presses the SEND key.

T_Calling party disconnect. Detected when an indication of the calling party disconnect has been received.

T_Network busy. Detected when the MSC has a resource unavailability for a call (e.g., TLDN, radio channel).

T_Radio busy. Detected when no radio channels are available.

T_Auth denied. Detected when the indication has been received that the handset is not authorized for call termination.

Call rejected. Detected when the handset informs the MSC that the call cannot be accepted.

T_No answer. Detected when no answer from the handset has been received.

6.3 Service Implications

If we compare existing WIN triggers with the AIN triggers and events completely specified at this time (nonshaded detection points), we will find that the majority of them are either the same or similar in both models. However, there are some important differences.

The first difference we already mentioned: lack of EDPs in WIN. This means that events cannot be installed dynamically from the SCP (HLR) when an application requires it. So, if an AIN application is built around a capability that relies on such dynamic installation of triggering points, it would not work the same way in the WIN environment. In addition, no notifications are available in WIN, which means that the applications that require only data collection based on occurring events would not be operating efficiently because all available WIN triggers suspend call processing when activated.

The second difference lies in the special needs for location searches and mobility in the wireless environment. The WIN call models contain a number of these special triggers that are not used in AIN: *O_Radio Busy, Location Request, T_Page*, and others. This is not surprising due to requirements mentioned before.

The third difference is a lack in the WIN terminating call model of two detection points: *Termination_Attempt* and *Term_Resource_Available*. Their unavailability affects the WIN service capabilities.

The *Termination_Attempt* trigger is contained within the T_NULL state—the initial terminating state. The activation of this trigger in the very beginning of the terminating call processing allows early intervention by an application residing in an SCP. Many services that depend on information retrieved from an SCP are based on this trigger activation. Among them are calling name delivery, selective call acceptance, do not disturb, and others. The general procedure is that when this trigger is encountered, a message is sent to the SCP, and the reply from the SCP carries information or instructions. Without this trigger services either are not supportable or are implemented in a less efficient fashion by using some of the triggers contained in the later call states.

The *Term_Resource_Available* trigger is contained within the SELECT_FACILITY state of the termination call model. This trigger (or event) can be activated when it has been determined that termination resources are available. Special treatment of the terminating line could be performed under the control of an application in an SCP. For example, selective ringing can be initiated based on the terminating user's profile stored in the SCP. Without this trigger such SCP-based services are not supportable.

Chapter

7

Applications

Applications are software programs that reside in computers and perform specific service-oriented tasks. Examples of applications are file transfer, money management, spread sheets, basic call processing, and call forwarding. We are more interested in the latter applications—call processing and others. However, all applications need to use some underlying capabilities, such as an operating system, database access, I/O handling, protocol handling, and others. For applications in telecommunications the protocols that applications use to communicate with their peers in remote network nodes are critically important. These protocols transfer information between remote systems, and applications expect functionality and reliability from them.

In general, three activities can be recognized in the area of applications: (1) development of applications, (2) development of application protocols, and (3) service creation. The first activity implies identifying and designing new applications to satisfy the needs of users. This activity is necessary in any networking and technological environment.

The last two items in the above list represent two different aspects of services that can be supported by wireless and wireline networks. Both are important in making intelligent network (IN)-based services transparent. The application protocol used in the network affects service functionality, and the service creation mechanism affects the introduction of new services.

7.1 Application Protocols

Application protocols enable the exchange of information among various network elements. The capabilities of application protocols limit the kind of services that networks can support. Two application protocols used for intersystem communications are important in our investigation: the Advanced Intelligent Network (AIN) application protocol and the IS-41 Mobile Application Part (MAP).

Protocols can be compared by looking at their protocol elements (messages and parameters) and procedures prescribed in specific circumstances. Similarity of protocols is key to providing seamless services. If one protocol is capable of transporting information that another protocol cannot transport, they cannot expect to support the same services in a seamless manner.

The above algorithm is workable when two defined protocols are being compared. Unfortunately, we need to make a big assumption about IS-41 becoming the WIN protocol. It is natural for us to assume this, and such a development is very realistic. And it is probably safe to say that some revision of IS-41 will support the wireless intelligent network (WIN), but it is impossible to say right now what it will mean: the addition of some messages and parameters or the creation of a different set of messages.

Similarities between IS-41 and AIN application protocols are easy to notice. Both protocols rely on Transaction Capabilities Application Part (TCAP) and therefore are developed in a similar Open System Interconnection (OSI) fashion. However, while the AIN application protocol has been specifically designed for IN operations, IS-41 has been initially designed for wireless operations without regard to IN services. The distributed nature of the wireless environment forced the IS-41 protocol to behave in the IN-like fashion, but it has been built with no detection points in mind.

In AIN, some messages are issued as the result of encountering a trigger or event. For example, the Info_Collected message is issued by an SSP when the Information_Collected trigger is detected. In the current revisions of IS-41 no such actions occur. Some parameters that are called *triggers* do exist, but the difference is that they are passed within messages to indicate what actually happened. For example, the TerminationTriggers parameter is passed in the LocationRequest response message. One can claim that regardless of the technique the information can be passed to and used by the application. That is true, but different approaches to the same problem do not help integration.

As was mentioned before, the fact that no WIN protocol formally exists prevents a comparison with the AIN protocol. The WIN document that contains the models refers to messages in some cases, but these messages do not always exist. Therefore, it is probably best to conclude that since the AIN protocol already exists, it would be beneficial to develop a WIN protocol in a way that creates significant similarities between AIN and WIN. This would be easier to accomplish if detection points and triggers were also similar (in the cases when it is plausible). What will actually happen remains to be seen, but this item will have a significant impact on the efficiency of a seamless network.

7.2 Service Creation

Service creation is the capability provided by INs that support a quick introduction of services within these networks by modifying existing functions in service control points (SCPs) (or similar network elements) or adding new ones. Service creation as it relates to IN is distinct from other service management issues, such as service activation. Service activation is an important element in the process of making seamless networks. If services are activated differently, there may be differences not only in management systems that handle this process but also in the services themselves because it is possible that system variations affect services. This would be covered in the configuration management part of the Telecommunications Management Network (TMN) (see Chap. 12).

Service creation, as we discussed it in Chap. 3, can also be accomplished using different implementations of service creation environments (SCEs). If this is the case, new services cannot be created in a seamless fashion across different platforms. Let's suppose that a telecommunications service provider maintains a network consisting of a wireless and a wireline portion. The wireline portion contains databases and an SCE built according to AIN, and the wireless portion contains an SCE built according to WIN. Unless specific care is exercised in making sure that the same features on two systems are accomplished the same way, the result will be a discontinuity in service seamlessness due to differences in service creation.

This is quite obvious and does not require a lot of proof. Unfortunately, there are no ready solutions for the potential discrepancies since neither environment, wireless nor wireline, has developed a standard in this area to which we can point. Since the SCEs for wireline networks were designed before the WIN work started, there is more practical experience in that area. It is probably safe to say that the same manufacturers would also build SCEs for wireless networks and, therefore, would assure continuity since it is often easier to modify a product than to build a new one. However, initial differences in services that may influence SCEs for the wireless environment may also create a significant difference in the future if discrepancies between the two types of SCE—for wireless and wireline networks—exist. Therefore, it is important for a network operator who wishes to create a seamless network to carefully choose the SCE that will be used in the network.

Chapter

8

Interconnection

If all functional and application protocol issues are resolved completely, can seamlessness be assured? Not if the underlying protocols are different. Even when networks of the same kind interoperate (or interwork), they have to be connected through physical links and protocol transport systems that both sides can handle. All layers of the protocol stack that we discussed in Chap. 3 have to interwork for the networks to be fully interoperational. We already discussed the possible differences between Advanced Intelligent Network (AIN) and Wireless Intelligent Network (WIN) application protocols in previous chapters. Now we need to focus on the rest of the protocol stack. Not surprisingly, this is the area of lesser concern from the seamlessness point of view because the same networking signaling protocol is used in both wireline and wireless environments. This protocol is Signaling System #7 (SS7).

There are two types of activities that SS7 supports: (1) call control activities and (2) transaction processing (non-call-associated activities). Call control functionality is provided by ISDN Services User Part (ISUP). Since cellular networks in the United States have been interconnected with wireline networks to set up calls for quite a while, we can say that this part of SS7 interconnection is working and should not present any problems.

Transaction processing is different. Transaction Capabilities Application Part (TCAP)—the part of SS7 used for transaction processing—carries both AIN and IS-41 messages. However, a seamless interconnection implies full compatibility and that is what we need to examine. In addition, differences exist between SS7 standards and implementations in the United States and the rest of the world. These differences have to be taken into account if a service provider is interested in building an international seamless network.

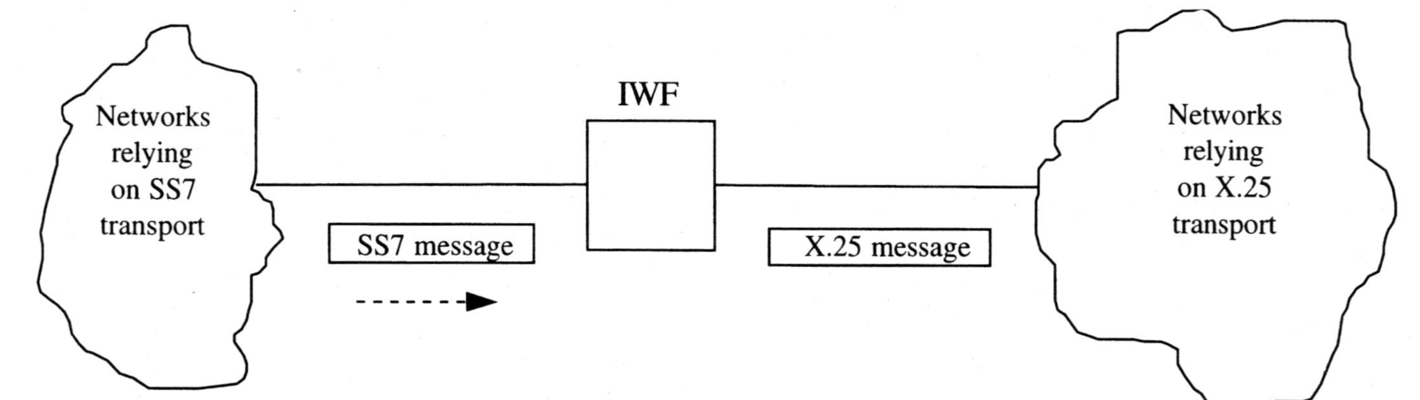

Figure 8.1 X.25/SS7 interworking.

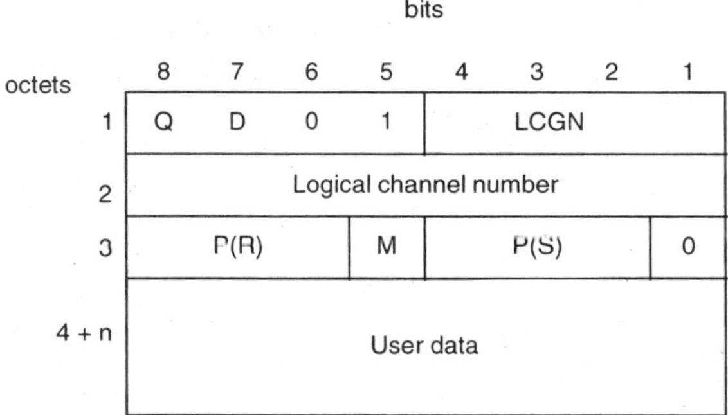

Figure 8.2 DATA message layout.

8.1 The U.S. Environment

Two issues that affect seamless network operations in the United States will be addressed here: the use of lower layers and global title translations (GTTs).

Lower layers

According to the IS-41 protocol reference model, TCAP can use either an underlying SS7 protocol stack or X.25 to transport information. The wireline networks primarily use SS7 as a transport mechanism. X.25 is used in many cellular networks. If a TCAP message has to cross a boundary between two networks, one using SS7 as a transport and the other using X.25, some interworking has to take place (see Fig. 8.1).

To get an idea of the differences between SS7 and X.25 transports we will compare a Message Transfer Part (MTP) message (Message Signal Unit, or MSU[1]) which is used to carry information between systems with an X.25 message used for the same purpose (DATA message). The X.25 DATA message format is shown in Fig. 8.2. The fields of the DATA packet have the following meaning:

Q-bit. Qualifier bit; it indicates whether user data or control information are contained in the packet.

D-bit. Delivery confirmation bit; it is used to indicate to the network that end-to-end acknowledgment for the packet is requested.

Bits 6 and 5, following the D-bit. When set to 0 1, indicate that packets for modulo 8 service are used.

---

1 MTP carries SS7 messages in what are called *signal units*.

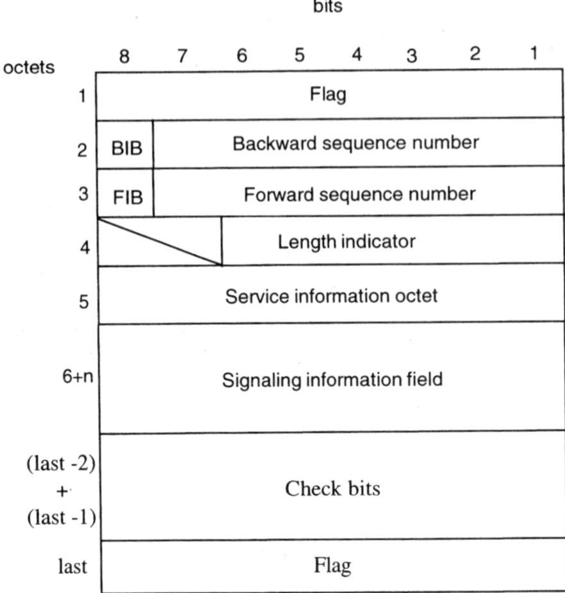

Figure 8.3 MSU layout.

LCGN. Logical channel group number; together with logical channel number identifies a logical channel assigned for the data call.

P(R). Packets received; it acknowledges the last received packet.

P(S). Packet sent; it identifies the number of the sent packet.

M-bit. More Data bit; it is used to indicate to the recipient of the packet that more data follows.

Bit 1 of the third octet. Always set to 0; it is the indication that the packet is DATA; no other packet type has this bit set to 0.

User data. User data are placed in this field, which can be up to 128 octets long.

The MTP MSU is shown in Fig. 8.3. The fields of the MSU have the following meanings:

Flag. Used to delimit the message

Backward sequence number (BSN). The sequence number of a signal unit being acknowledged

Backward indicator bit (BIB). Used in the error control method

Forward sequence number (FSN). The sequence number of the signal unit in which it is carried

Forward indicator bit (FIB). Used in the error control method, together with BIB, BSN, and FSN

Length indicator (LI). Indicates the number of octets following this field and preceding the check bits

Service information octet (SIO). Used to indicate the SS7 user part that issued the message (e.g., ISUP, TCAP)

Signaling information field (SIF). Carries all necessary data and is at least 2 and at most 272 octets long

Check bits (CK). Error detection bits

The two messages we are comparing carry signaling information between network nodes. This information would be placed in the *User data* field of the X.25 message or SIF of the SS7 message. The interworking function mapping X.25 and SS7 transport protocols would have to translate all applicable parameters from one protocol to another. The difference in data field sizes would also have to be accommodated, possibly by segmentation of larger messages and subsequent reassembly.

Global title translation

GTT is the process of mapping some information, called *global title*, to an address associated with a particular node in the network. The process of mapping is based on the code specifying the translation to be performed. In short, a GTT type points to a particular mapping table. A number of translation types have been standardized by ANSI (see standard T1.112). Table 8.1 lists these translation types. The issue that can affect the process of interconnecting various networks is whether existing types are sufficient to support required services. If the current translation types are not adequate for a particular service, new ones would have to be added via the standardization process. Examples of the need for additional translation types could be new services that require mapping of new global titles to destination addresses. For example, if a service is based on the names input by subscribers, and these names would have to be used as the indicators of where the necessary database is located, a new translation type may have to be added to address this need.

8.2 ANSI versus ITU-T

So far, the discussion has focused on the possible issues if all interconnected networks are within U.S. boundaries. If international net-

Table 8.1 ANSI Standardized Global Title Translation Types

Application group	Binary	Decimal
Reserved	00000000	0
Identification Cards	00000001	1
14-Digit Calling Card	00000010	2
Cellular Nationwide Roaming Service	00000011	3
Global = Point Code	00000100	4
Calling Name Delivery	00000101	5
Call Management Application	00000110	6
Message Waiting	00000111	7
SCP-Assisted Call Processing Applications	00001000	8
Internetwork Applications	00001001- 00011111	9 - 31
Spare	00100000- 10111111	32 - 191
Network-Specific Applications	11000000- 11111010	192 - 250
Call Management Application	11111011	251
Network-Specific Applications	11111100	252
14-Digit Calling Card	11111101	253
800 LIDB	11111110	254
Reserved	11111111	255

working is desired, the differences between the SS7 standards defined by ANSI and ITU-T have to be taken into account. And there are differences between them. The usual way of dealing with the situation is to use gateways between countries to perform the functions of interworking units—translating one type of a protocol to another. This is probably the solution that will stay with us for a while, but it would be useful to understand the existing differences anyway. One reason for this is a possible requirement of a seamless network operator to decrease processing time taken by the gateway functions, and another might be an inability to provide some services if certain differences exist.

The gateway solution works today for ISUP procedures. Although there are differences in ISUP implementations in different countries, the nature of the call processing requires the gateway switches to be involved with every call setup. While at the gateway, ISUP messages adapt the look expected at the other side of the gateway switch. For example, a 24-bit point code identifying an SS7 node in the United States is changed into the 14-bit point code used in Europe.

TCAP interworking presents a different problem. Figure 8.4 shows two international interworking scenarios: for call control and for transaction processing. In the case of call control, an intermediate switch has to be involved because it has to connect a U.S. trunk with an international one. In the case of transaction processing the goal would be to access the desired destination database directly from a gateway STP. This is because all that is needed from the transaction is to exchange information between an originating node and an SCP in the foreign country. It would be better not to impose a mapping process on the STP because it not only delays processing but also makes the STP very complex. This complexity, of course, is the result of differences between ANSI and ITU-T standardized SS7 parts. The most important differences are in TCAP and SCCP, and the presumed gateway function would have to deal with both.

8.2.1 SCCP

Some differences exist between the Signaling Connecting Control Part (SCCP) implementations based on ANSI and ITU-T standards. The first is a parameter which is currently a part of the U.S. standard only—Intermediate Signaling Network Identification (ISNI). This parameter is needed to specify which intermediate signaling network is to be used for SS7 message routing. ISNI is not available in the ITU-T standard, and, therefore, non-ANSI-compliant SS7 equipment would not be able to deal with this parameter.

Another difference is in the coding of Calling/Called party address parameters. While both standards use the same elements within the parameter, their sequence is different. A generic structure of the Calling/Called party address parameter is shown in Fig. 8.5 and is the same in both standards. The differences between ANSI and ITU-T standards exist in both components of this parameter. Figure 8.6 shows the differences in coding of ANSI and ITU-T Address indicator, and Fig. 8.7 shows the differences in coding of the Address elements. In the Address indicator, the SSN Indicator and Point Code Indicator are reversed. In the Address, the sequence of Signaling Point Code and Subsystem Number are also reversed.

One more difference is related to implementations, not standards. Both ANSI and ITU-T standardize connectionless and connection-oriented classes of SCCP service. However, the connection-oriented option has not been implemented in the U.S. signaling networks. In the scenario of international interworking (as in Fig. 8.4) a message coming into the U.S. network from overseas which uses a connection-oriented class of service would not be understood, and a failure would be reported. Another problem situation would be experienced by a manufacturer wanting to use in the United States products built for, let's say, a European market. Such a product would not be able to operate in the existing U.S. signaling networks if it relied on the con-

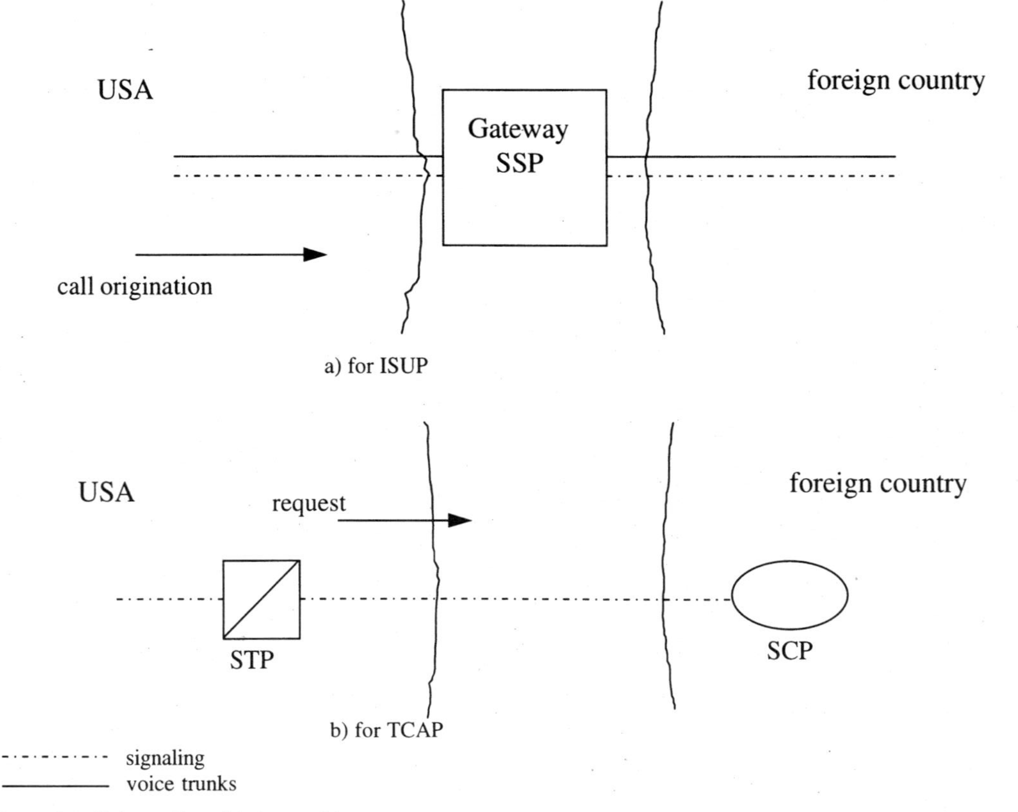

Figure 8.4 International interworking.

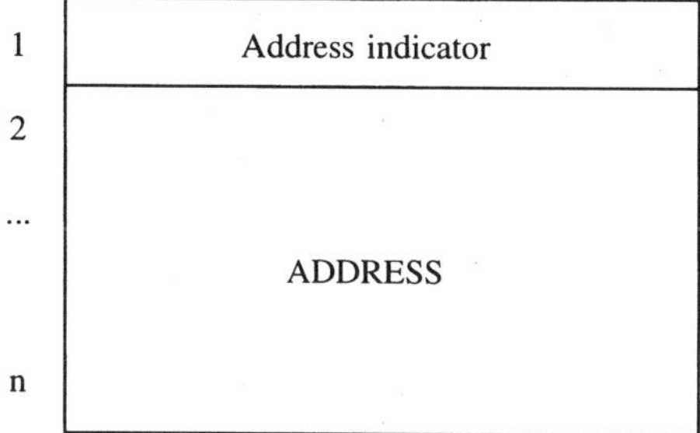

Figure 8.5 Calling/called party address.

nection-oriented class of SCCP service. A modification of this product would be required.

8.2.2 TCAP

Several discrepancies exist between the ANSI and ITU-T TCAP standards. Interworking of TCAP messages at the international boundary requires mapping of different parameters if discrepancies are not resolved. The discrepancies are:

1. *Package types.* In the ANSI standard among package types for messages there exist types with and without *Permission.* It applies to Query and Conversation package types, and the ANSI standard specifies them with and without permission. No such options exist in the ITU-T standard.[2]

2. *The use of Transaction IDs.* While both standards support the use of two transaction IDs (Originating and Responding), their encoding is different. To identify the difference we need to explain the principle of parameter encoding. Each parameter in a message is encoded using the scheme shown in Fig. 8.8. According to this scheme a parameter has a type, which identifies its function, length, and content. In the ANSI standard both transaction IDs (originating and responding) are included in the same parameter;

---

[2] There is also a difference between ANSI and ITU-T terminology. While the ANSI standard refers to package types, the ITU-T standard refers to message types. And some message types have different names. Compare ANSI package type names introduced in Chap. 3 with the following ITU-T message type names: Unidirectional, Begin, End, Continue, Abort. Functionally, *Begin = Query, End = Response, Continue = Conversation.*

bits

8	7	6	5	4	3	2	1
Nat. Ind.	Rtg. Ind.	Global title indicator				SSN Ind.	Point Code Ind.

a) ITU-T coding

bits

8	7	6	5	4	3	2	1
Nat. Ind.	Rtg. Ind.	Global title indicator				Point Code Ind.	SSN Ind.

b) ANSI coding

Nat. Ind. -- National indicator
Rtg. Ind. -- Routing indicator
SSN -- Subsystem Number

Figure 8.6 Comparison of address indicators.

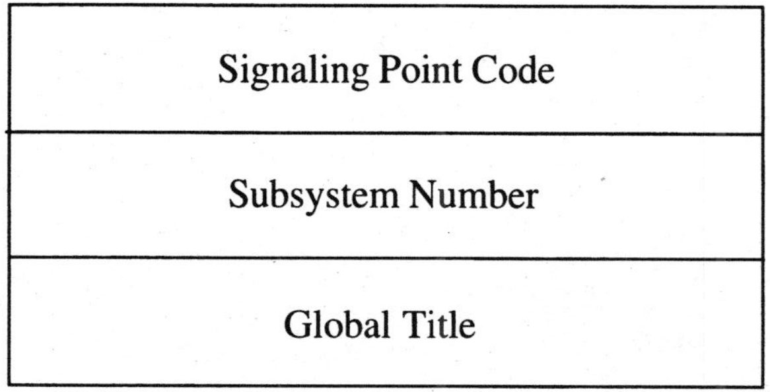

Signaling Point Code
Subsystem Number
Global Title

a) ITU-T coding

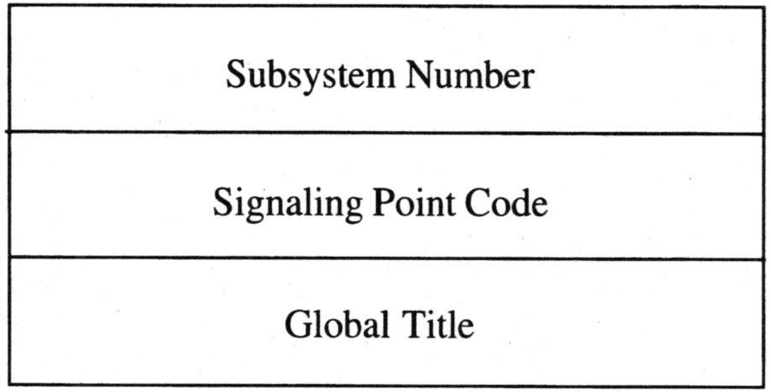

Subsystem Number
Signaling Point Code
Global Title

b) ANSI coding

Figure 8.7 Comparison of address elements.

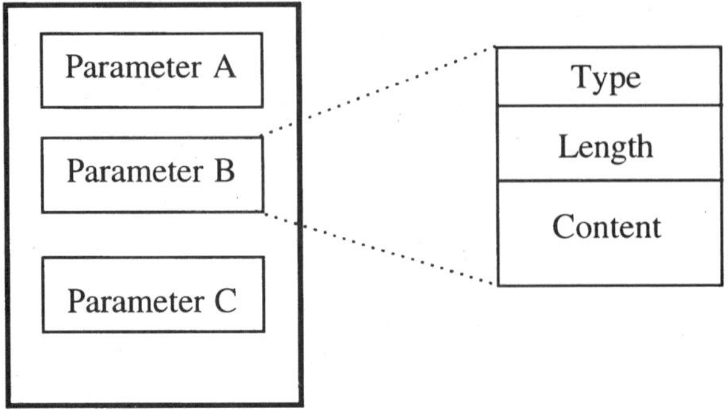

Figure 8.8 Parameter encoding.

in the ITU-T standard two different parameters are used (see Fig. 8.9). The result is that, although the information is the same, the parsing algorithms in the two cases are different.

3. *Component types.* While in the ANSI standard six component types exist (see Chap. 3), ITU-T specifies five component types. The Invoke component in the ANSI standard can be of two types: *last* and *not last.* The ITU-T standard does not include this differentiation.

4. *Dialog portion.* The dialog portion has been added to TCAP standards relatively recently. It provides association control and security functions. The difference between the ANSI and ITU-T standards lies in their reliance on another ITU-T standard—Association Control Service Element (ACSE). While the ITU-T standard complies with ACSE services, the ANSI standard is not as compliant. The result is that the ANSI dialog portion of a TCAP message requires less space than a comparable ITU-T message.

Type
Length
Originating ID
Type
Length
Responding ID

a) ITU-T encoding

Type
Length
Originating ID Responding ID

b) ANSI encoding

Figure 8.9 Different encoding of two transaction IDs.

Chapter

9

Seamless Network Architecture

Network architecture is the foundation for all the services that a service provider offers. Architectural decisions affect the network's ability to support services at an appropriate performance level. A possible evolution toward future services and capabilities also depends on the type of the architecture selected. A properly built network architecture allows the flexibility and the freedom to deploy different services, perform at the required level, and migrate into a new environment when necessary.

Architectural decisions are affected by the circumstances of a particular service provider: whether the service provider is building a new network, extending an existing one, or requires some combination of both. Wireline and cellular networks exist today, and their operators need to evaluate the needs for seamlessness against the network architectures they already have. Possible additions and modifications to the existing architectures would be based on these needs. Personal Communications Services (PCS) networks are currently being built. This process will be different for different service providers, but it would have to be either a completely new network or a combination of some existing and some new network equipment. Building a new network allows more freedom in selecting the architecture. However, even a completely new network would have to interwork with other networks, and this interworking has to be taken into account. When a decision on the network architecture is made, it should always be made with future needs in mind, and seamlessness should be one of the goals.

In this chapter we will address the architectural issues primarily from the functional point of view. Building a physical network requires knowledge of the specific conditions, traffic, and geography.

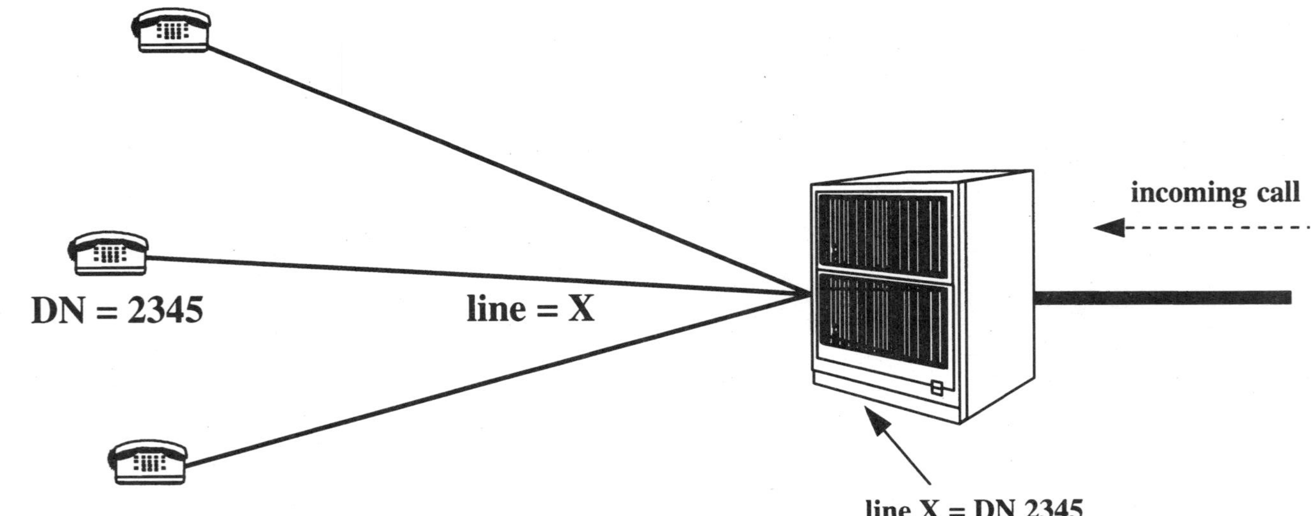

Figure 9.1 Identification of terminals in wireline networks.

The functional network architecture provides the basis for the physical network and therefore is sufficient for our discussion. Furthermore, we will focus on the intelligent network (IN)-related aspects of the network architecture because that is where the majority of service and evolution-related issues occur.

9.1 Convergence Process

The process of convergence requires careful analysis of different components of the networks not only in terms of the protocols they use and functions they perform but also in terms of the implementation details of these functions. Variations in service environments sometimes translate into variations in implementation of the same functions. To demonstrate this point let's look at switches deployed in wireline and wireless networks.

Switches in both networks perform very similar functions. Their most profound differences are primarily in the access interfaces they serve. Wireline advanced intelligent network (AIN) switches serve POTS and ISDN interfaces. Mobile switching centers (MSCs) serve interfaces to radio equipment (base stations and base station controllers). This difference generates functional discrepancies. When we discuss services that networks can provide to their customers, some of the limitations come from the limitations of the network access interfaces. If an interface does not handle a particular service because it is not capable of delivering some data to the end-user device, the capabilities of the rest of the network do not matter. The access interface becomes a limiting factor.

The wireline switch, a switching service point (SSP), connects "wired" users and identifies these users by a directory number (DN). A DN is mapped onto a particular subscriber line and, when this particular subscriber is called, the switch examines the called number (destination address), finds that this number is associated with a particular line, and initiates alerting on that line (see Fig. 9.1).

The wireless switch, an MSC, has different connectivity to subscribers. Since the subscribers are not "wired" to the switch, they move around—roam. At any particular moment a different set of subscribers is served by a wireless switch. The numbers by which these subscribers are identified and addressed (the Mobile Identification Numbers—MINs) are not associated with any specific lines that the switch supports. Figure 9.2 shows connectivity of subscribers to an MSC.

When an MSC attempts to deliver a call to a subscriber with MIN = 2345, it has to associate this number with a particular line or a chan-

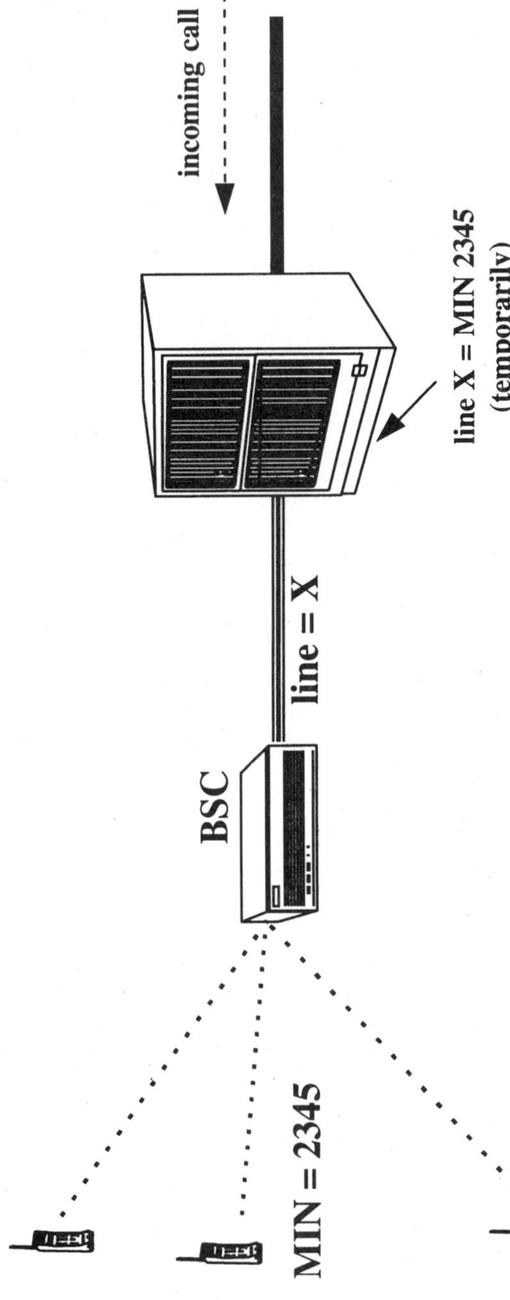

Figure 9.2 Identification of terminals in wireless networks.

nel that provides communications between an MSC and the subscriber through radio equipment (base station and base station controller). To accomplish this, a temporary DN is assigned for a particular call for as long as it is needed to create an association (connection). Then this temporary number is reused. It is called a temporary line DN (TLDN). A TLDN allows a temporary association between a physical switch access line and a wireless subscriber to be created. This process is considerably different from the process in a wireline switch because of the need for a temporary association.

As we see, even an easily understood switch function of associating subscribers' numbers with switch termination facilities is performed differently due to differences in the two environments. The task of defining architecture alternatives for a seamless network involves deciding how to configure wireless and wireline systems so they can work together and provide seamless services. The network functions that can be manipulated for placement in a seamless network are SSPs, MSCs, service control points (SCPs), home location registers (HLRs), visitor location registers (VLRs), service nodes (SNs), and intelligent peripherals (IPs). And let's remember that the driving forces behind the integration process are oriented toward services and operational savings. That is, the attempt is to provide an environment which supports seamless services less expensively.

9.2 Architectural Alternatives

The object of a network architecture is to define how various network components are interconnected to support a given set of services. For the task at hand—integration of wireless and wireline networks—a very high-level model of a seamless network could be as represented in Fig. 9.3. Now we can start identifying network architecture alternatives that can be applied to the goal of seamless networks. Note that radio equipment [base station controllers (BSCs), and base stations (BSs)] is not shown since the intent is to identify networking issues; radio is considered to be part of the access. Operations systems are also not shown because, assuming Telecommunications Management Network (TMN) implementation, operation systems' (OSs') connectivity with other network elements would look, in principle, the same (see Chap. 12 for more details).

9.2.1 Alternative 1

Figure 9.4 shows a network architecture which is built out of two distinct components: wireless and wireline. The MSC and associated HLR and IP constitute the wireless component, and the SSP with

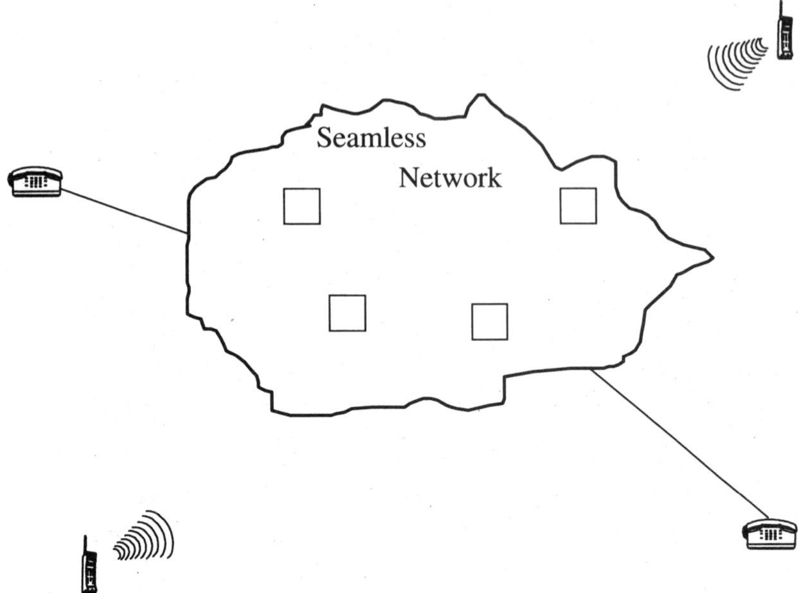

Figure 9.3 General model of a seamless network.

associated SCP and IP constitute a wireline component. In the wireless area, the IP supports services which are IN-like. For example, it can support specialized applications, such as voice-activated dialing. The HLR performs the usual functions of subscriber information storage, subscriber location tracking, etc. (a VLR is assumed to be collocated with the MSC). The application protocols used between various wireless network elements are as follows:

MSC—HLR: IS-41, including WIN additions

MSC—IP: A specialized protocol based on existing standards, such as DSS1, IS-41, etc.

In the wireline area, the SCP supports AIN functions and the IP supports specialized applications which can also be voice-activated dialing. The application protocol used between an SSP and an SCP is the AIN application protocol; the interface between an SSP and an IP is also specialized and could be based on Digital Subscriber System 1 (DSS1), AIN, etc.

All application protocols in this architecture (with the possible exception of a protocol supporting an IP) use Signaling System #7 (SS7) as a transport. This is represented in the architecture by the signal transfer point (STP) placed at the center of a network. All nodes

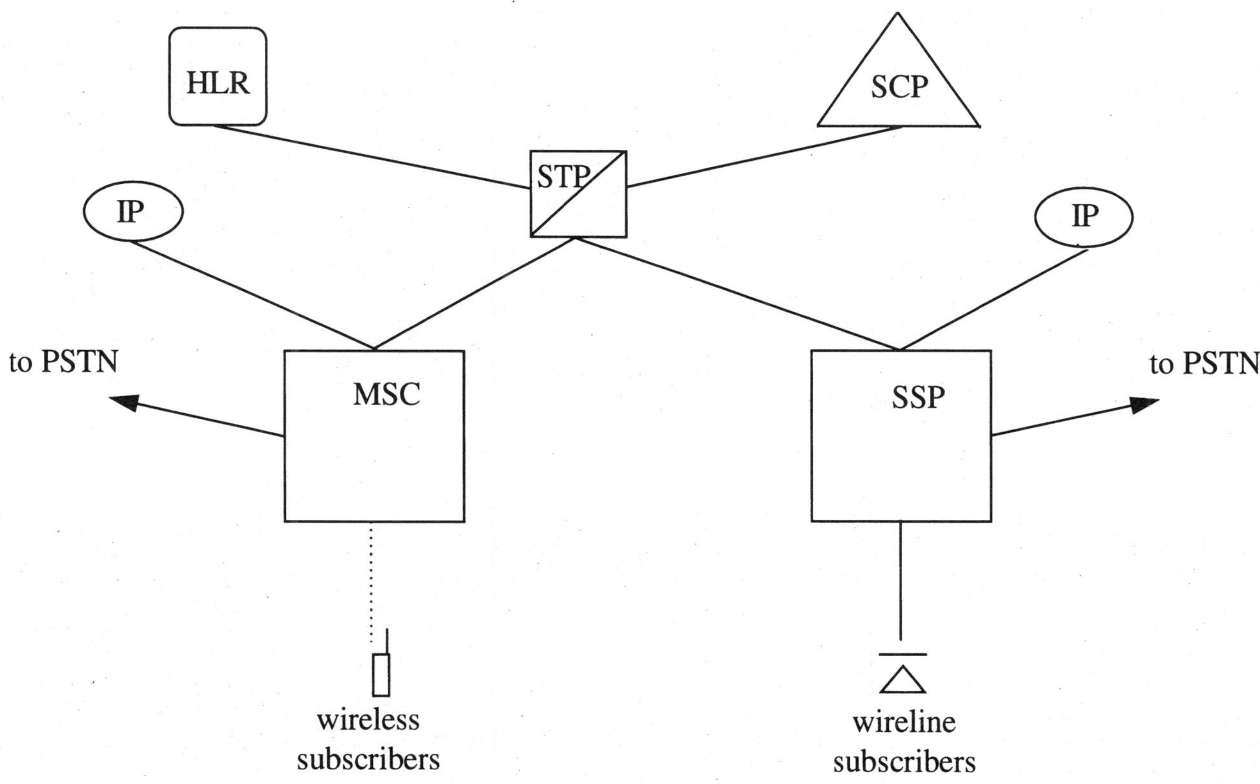

Figure 9.4 Alternative 1.

on this SS7 network have point codes and could be accessible to all other nodes. The common transport mechanism provides a degree of seamlessness by simplifying the interconnection. The interfaces to the IP elements depend on their functionality and other factors, such as availability of suitable equipment. If an SS7-based protocol is used for this purpose, the IP would be connected to the STP instead of a switch.

A potential benefit of this arrangement is the possibility to access databases which were initially designed to support either wireless or wireline components. For example, if an HLR is a system that stores calling names of wireless users and an SCP is the same system for wireline users, they can be cross-referenced by the switches serving the opposite sides of the network: an SSP could access the HLR for this purpose and an MSC could access the SCP. Obviously such a cross-reference would require the systems to be able to recognize different protocols, in this case IS-41 and the AIN application protocol.

This network architecture, although providing a low level of integration, does provide a platform for services which are common to both wireless and wireline subscribers. It is just a platform because connecting equipment as shown in Fig. 9.4 is not a guarantee that service commonality, (or seamless services) can be supported. Unless it is a totally new network built with newly designed equipment, additional efforts will be necessary to align services and to provide them in a seamless fashion.

9.2.2 Alternative 2

This alternative (see Fig. 9.5) focuses on common switching; that is, a switching system performs the functions of both the SSP and the MSC. It implies that the VLR function is collocated with the MSC. Commonality of switching translates into commonality of services, provided the nonswitching parts of a service supported by other network elements are also common. But, many functions that are required from a network to support a typical service are switch-related because the switch is a handler of call processing. In this arrangement a switch would have to be capable of handling two call models: wireline and wireless. This probably implies a need to integrate the two models and create a superset representing both environments. The call states of this supercall model that were not applicable in a particular environment would have to be bypassed. For example, calls originated by a wireline subscriber would bypass call states associated with radio operations because they are unnecessary in the wireline environment.

Assuming that all necessary call model differences are resolved, this architecture provides a means for common services and for common

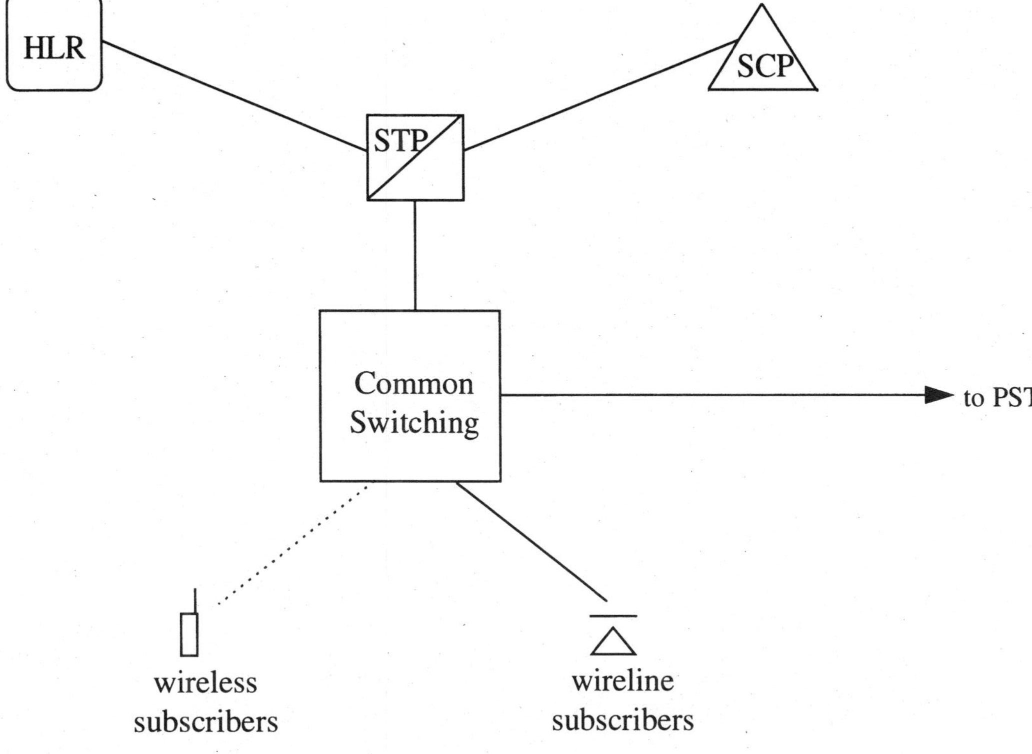

Figure 9.5 Alternative 2.

database access using the same or different protocols. This alternative implies different sets of databases for mobility and service needs of a network. Mobility information resides in the HLR, and IN service-related information resides in the SCP. The SCP, therefore, would serve the needs of both wireless and wireline subscribers. Depending on the distribution of information between the HLR and SCP, there may need to be a communication link between them. Such a link, as well as the switches to HLR and SCP interfaces, are supported by the SS7 signaling network.

The HLR is accessed via the IS-41 protocol, while the SCP is accessed via the AIN application protocol. This translates into the need for a switch to be capable of using both of these applications protocols. The application protocol duality requires a switch to discriminate events and decide which protocol to use depending on the targeted database. This is added complexity, but it allows the usage of existing databases without any modifications and provides an evolutionary path forward. An exception is the SCP-to-HLR application protocol. If required, this protocol will introduce a new requirement for either an HLR or an SCP. If this protocol is AIN-based, the HLR would have to adapt to the new interface, and if the protocol is IS-41-based, the SCP would have to change.

Figure 9.6 shows an example of call flows with a common switching system when wireless and wireline users originate calls. In this figure "terminals" refers to both wireline and wireless terminals. Activities related to a wireline user are indicated with a solid line, and activities indicated with a broken line are related to a wireless user. The switch handling both types of users differentiates between them and issues different messages.

a) A wireless user originates a call; this user requires authorization before being served, and an AUTHREQ message is issued to the HLR.

b) A wireline user originates a call at about the same time as the wireless user; an Off-Hook Immediate trigger is encountered, and the Origination_Attempt message is sent to the SCP.

c) A response from the HLR authorizes the wireless user to make a call (authreq message).

d) An SCP response is contained in the Analyze_Route message.

e) The wireless user subscribes to an IN service which requires a communication with the SCP; a made-up SERVREQ message is sent to the SCP according to this service's needs.

f) The Info_Analyzed message is sent to the SCP as the result of encountering the Info_Analyzed trigger.

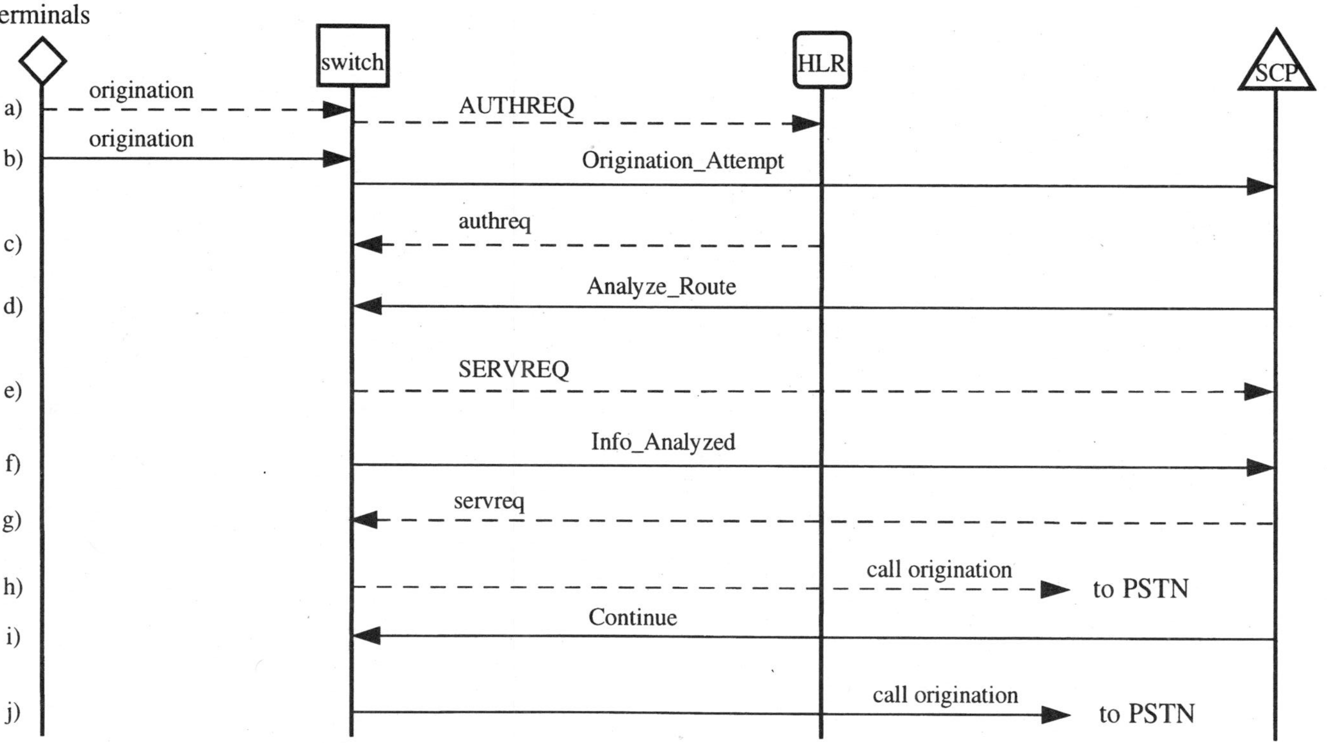

Figure 9.6 Example call flows for alternative 2.

g) A response from the SCP is received in the servreq message.

h) A call originated by the wireless subscriber can now be routed to the public switched telephone network (PSTN).

i) A response from the SCP is contained in the Continue message directing the service switching point (SSP) to continue with the call.

j) The call originated by the wireline subscriber is routed to the PSTN.

The advantages of a common switch approach, in addition to common services, are common billing arrangement, common provisioning, and common evolution, the last one being extremely important because of the high costs typically associated with switch upgrades. The disadvantages are growing switch complexity and the negative impact of this complexity on switch performance.

9.2.3 Alternative 3

While alternative 2 focuses on common switching, alternative 3 focuses on the commonality of databases. For this alternative we will maintain the same two databases as in the previous alternative (see Fig. 9.7). Two distinct switching systems rely on the common database platform which houses both the HLR and the SCP. In this architecture the VLR is still assumed to be collocated with the MSC. SS7 is used as the transport for the application protocols, and two cases can be envisioned for database access: (1) IS-41 and the AIN application protocol and (2) a common protocol.

A common application protocol in the architecture of alternative 3 would simplify many issues. Whether it is IS-41 encompassing AIN functionality or the AIN application protocol encompassing mobility requirements, the end result would be a single protocol—a single language—for communicating various needs to appropriate applications. It is possible to develop a protocol that contains a sufficient variety of elements to support the requirements of both environments. This protocol would work by accessing particular applications through an association mechanism such as the dialogue portion of the Transaction Capabilities Application Part (TCAP). Such a common protocol is not available right now, but IS-41 may be moving in the direction of providing support for more of the IN-like functions and thereby becoming a common protocol in the future.

In the meantime, case 1 for accessing a common database is more plausible because it does not require protocol modifications. Instead, it requires a platform capable of handling both protocols (speaking

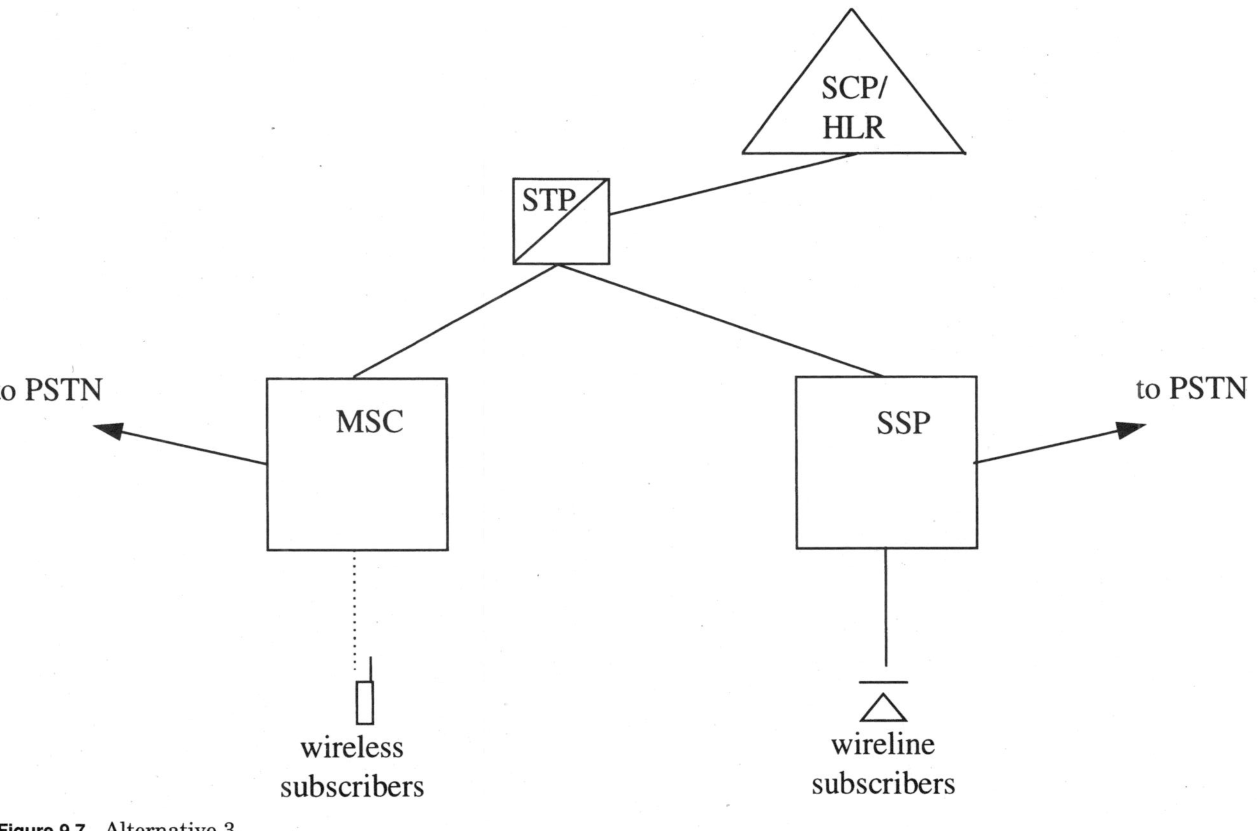

Figure 9.7 Alternative 3.

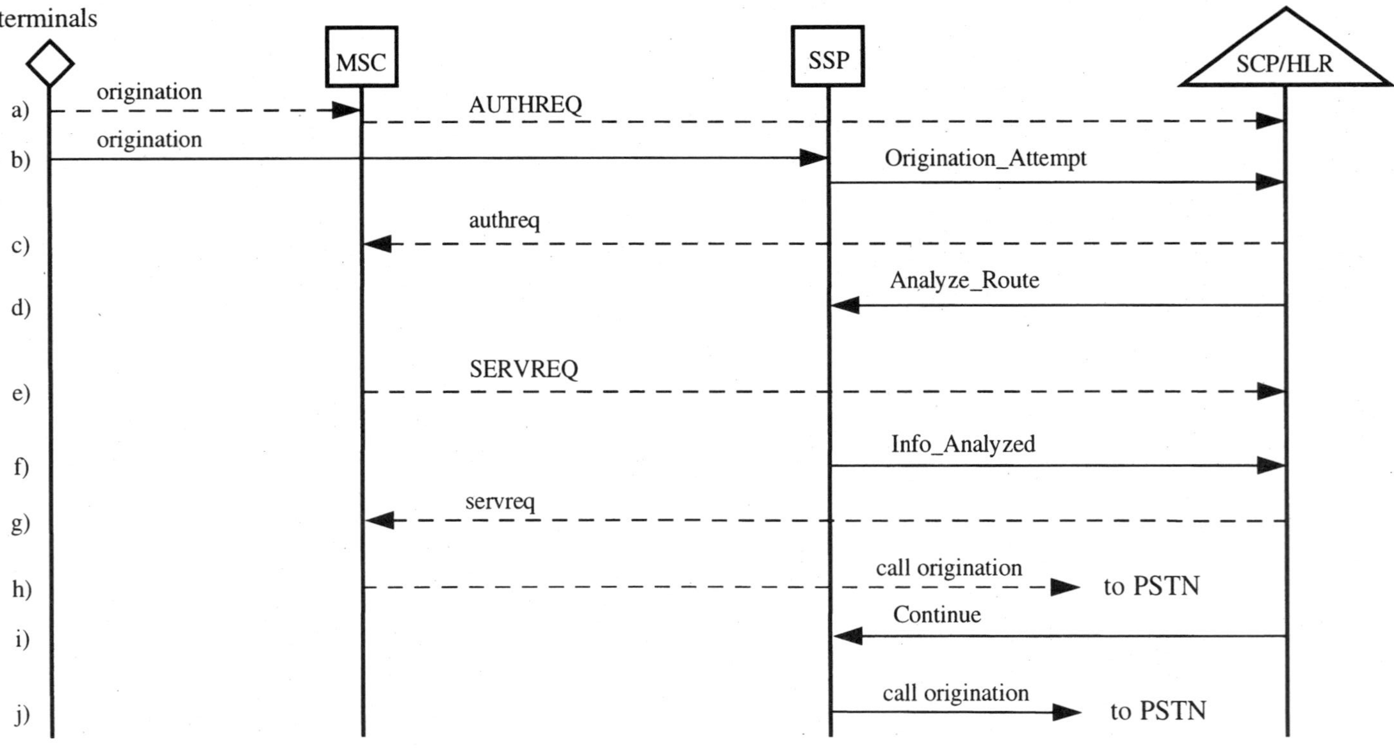

Figure 9.8 Example call flows for alternative 3.

two languages, so to say). This kind of a system is possible to develop, and there are already some available.

Figure 9.8 shows the call flows for the situation discussed in alternative 3. With separate switching systems origination of messages is different, but the target system for these messages is the same. A common application protocol would simplify the operations of the common database platform but would require one or both of the switching systems to modify existing operations.

A common database, which is a common service platform, can also be very beneficial in terms of service evolution. Service requirements may dictate new capabilities that could or could not be developed using existing platforms. The next alternative addresses greater service functionality which might be desirable.

9.2.4 Alternative 4

This alternative builds on the previous one and adds a new element which is also contained within a common database platform. This element is a service node, or SN (see Fig. 9.9). The functionality of an SN is not defined in the way other network elements are. According to some industry participants, an SN is a combination of an SCP and IP, but this view is not shared by many. Because of this lack of definition, it is difficult to find a place for SNs in the architecture without making some assumptions about their functionality.

Lack of formal definition of both an SN and IP could be replaced by a de facto definition of actual products. However, scanning the ideas that different industry members put forward for their implementations of SNs and IPs does not show any uniformity among them. The result is that the roles of these functional elements are different depending on the manufacturer. This is probably the way it will be because different manufacturers develop what they think are competitive features in these products and differentiate themselves by emphasizing one or another aspect of an SN or IP.

For the purposes of this book an SN is a set of functions distinct from those provided by an SCP. It handles specialized applications and potentially can be equipped with a service creation environment as an SCP. It may contain functionality associated with circuit or packet-switching, which indicates that there are similarities with IPs. As we discussed before, an IP is also an undefined element but is one which contains some specialized logic and hardware, such as announcement equipment. We will define an IP to be a simpler version of an SN, an entity that has specialized functions but not a lot of capabilities in terms of creating and deploying many new services, something that an SCP or an SN has.

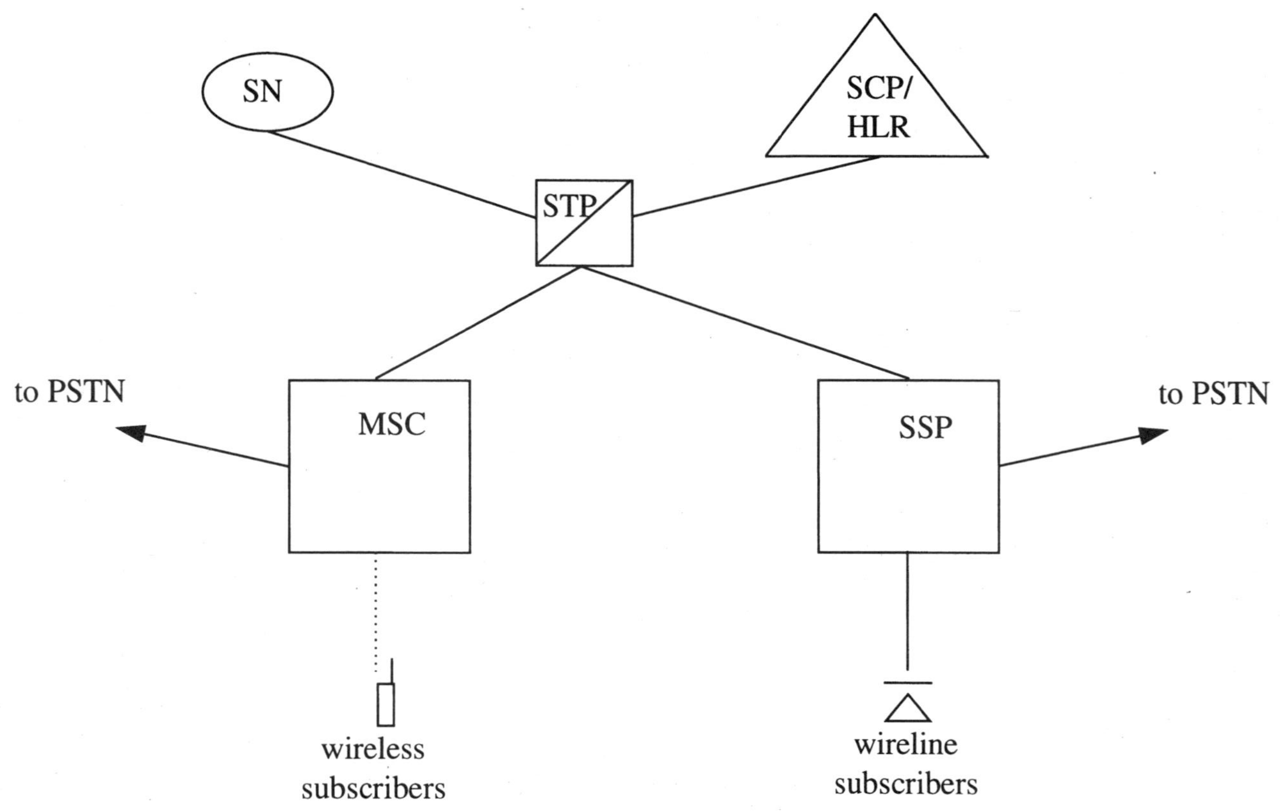

Figure 9.9 Alternative 4.

With this caveat, what does it mean to add SN functionality to the existing platform containing an HLR and an SCP? It means expanding service capabilities of a network by adding a specific identifiable set of functions. These functions could potentially be implemented in an SCP or HLR, but some specific requirements create the need for an independent unit—an SN or an IP. The requirements affecting the need for an SN or IP include the following:

- *Performance targets.* A prospective system would achieve better performance if it contained only a specialized set of services, distinct from those supported by existing network capabilities in the SCPs and HLRs.
- *Geographical distribution.* As with any network design, concentration of subscribers in particular areas requires service support in those areas and may require a specialized database to be located at a particular place.
- *Security considerations.* If some information requires more security precautions than others, a higher level of security could be provided in a particular system instead of in all of them.
- *Specialized focused features.* It may be beneficial to develop a standalone node supporting only specific features; this is usually the reason for using an IP.
- *Savings.* If projected traffic is relatively small in a startup network or in a particular part of a network, it may be possible to save resources by combining many of the SCP and other functions together and create an SN, even though later in the network development a fully developed individual set of functions (HLR, SCP) will be deployed.

This alternative does not explicitly show an IP; as we have defined an SN, it can contain specialized IP functions. It is also possible to include separate SNs and IPs in the network because of traffic and service considerations. Figure 9.10 shows such a network architecture with two IPs, each performing targeted functions for both types of switches. For example, one IP could support voice-activated dialing, while another supports voice messaging. They are shown to be connected to switches implying non-SS7-based signaling. However, SS7 could also be used if the IP equipment had such an interface.

9.2.5 Alternative 5

The next logical step in the evolution of architectural alternatives is to add an SN to the existing common database platform. Figure 9.11

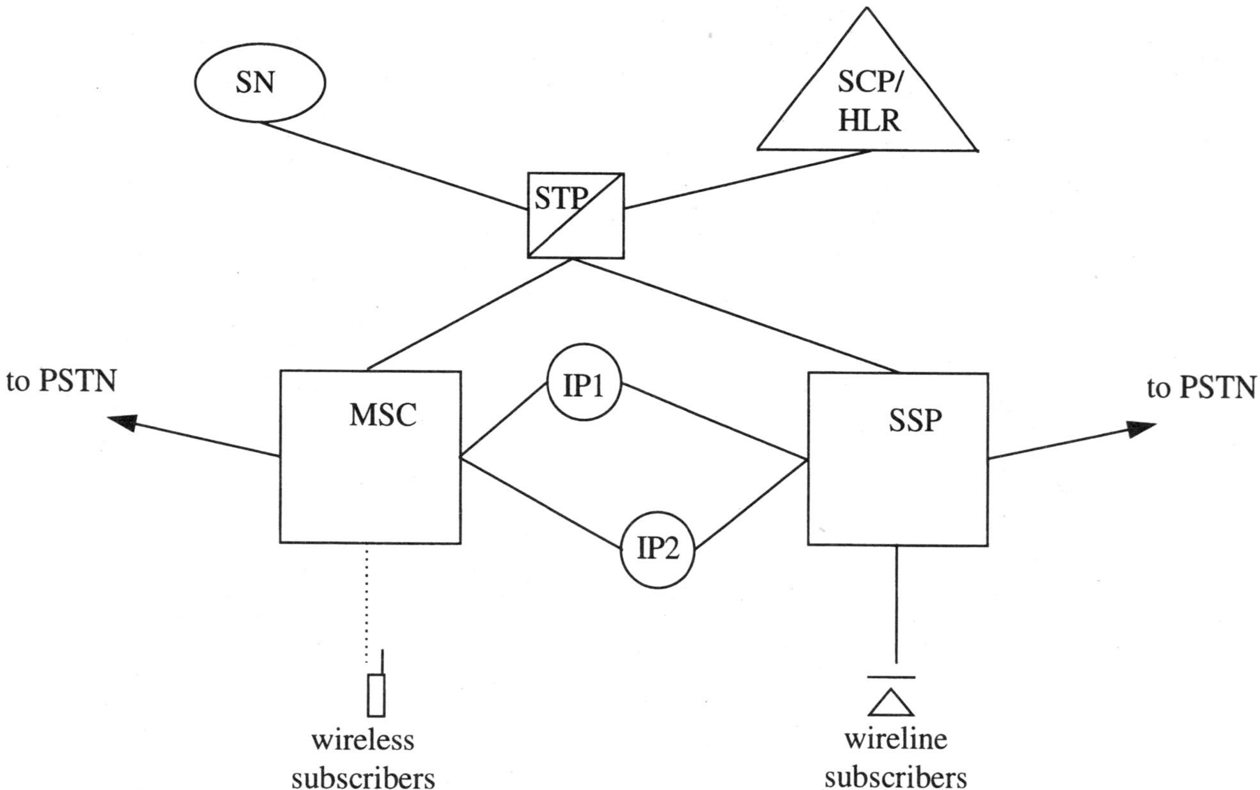

Figure 9.10 Alternative 4 with IPs.

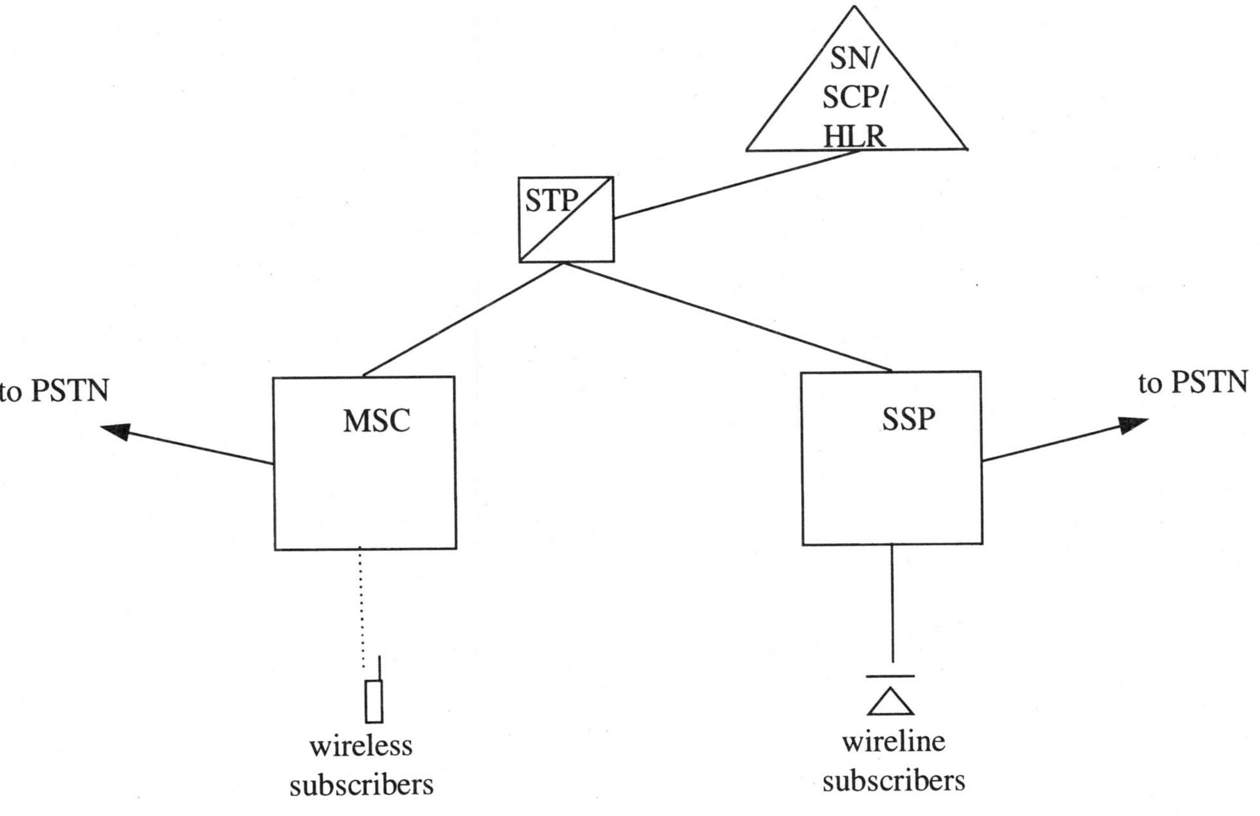

Figure 9.11 Alternative 5.

shows the alternative that combines all possible functional sets in the same system. The same observations as for alternative 3 can be made regarding the application protocols in this architecture.

It is interesting to compare scenarios with different protocol approaches and see how they would affect the architecture of the network. As an example, if SN and SCP applications use the same protocol—the AIN application protocol—it is more natural to collocate an SN with an SCP and have an HLR in a separate system. If all the applications supported by an SCP, SN, or HLR use different application protocols, it may be beneficial to locate all of them separately. That would really take us back to alternative 1, where we discussed merging wireless and wireline networks. If all the protocols used by applications in the network are different and all applications are located in different systems while switching is not common, the degree of integration achieved in such an architecture depends on the degree of similarities between services provided to wireless and wireline subscribers. On the other hand, a common database platform would support seamless services and their common evolution. The limiting factor for a seamless network evolution in this case is noncommon switching.

A variation of alternative 5 is presented in Fig. 9.12. This architecture follows a possible functional definition of an SN as a combination of an SCP and an IP. In this case an SN performs SCP functions in addition to specialized services and, therefore, handles the AIN application protocol. Actually, all architectural alternatives with SNs discussed here can exclude an SCP (this is similar to Fig. 9.12). We will not reiterate this possibility. The important point is to recognize that a combination of SNs, SCPs, and IPs as a collection of functions can be different depending on service requirements, network topology, and other considerations.

9.2.6 Alternative 6

In the ideal world the architecture of a seamless network would be quite simple (see Fig. 9.13). A common switching system, capable of serving wireless and wireline subscribers, is connected to a common database platform housing all necessary types of applications. This architecture does not imply the star configuration for the database access; remember, this is a functional architecture, not a physical topology. When performance of a single system is not sufficient to handle network traffic, more of the same common systems would be deployed. The fact that they are common allows the use of any of the multiple systems to be flexible and depend on particular network requirements. All of these systems would support the same interfaces

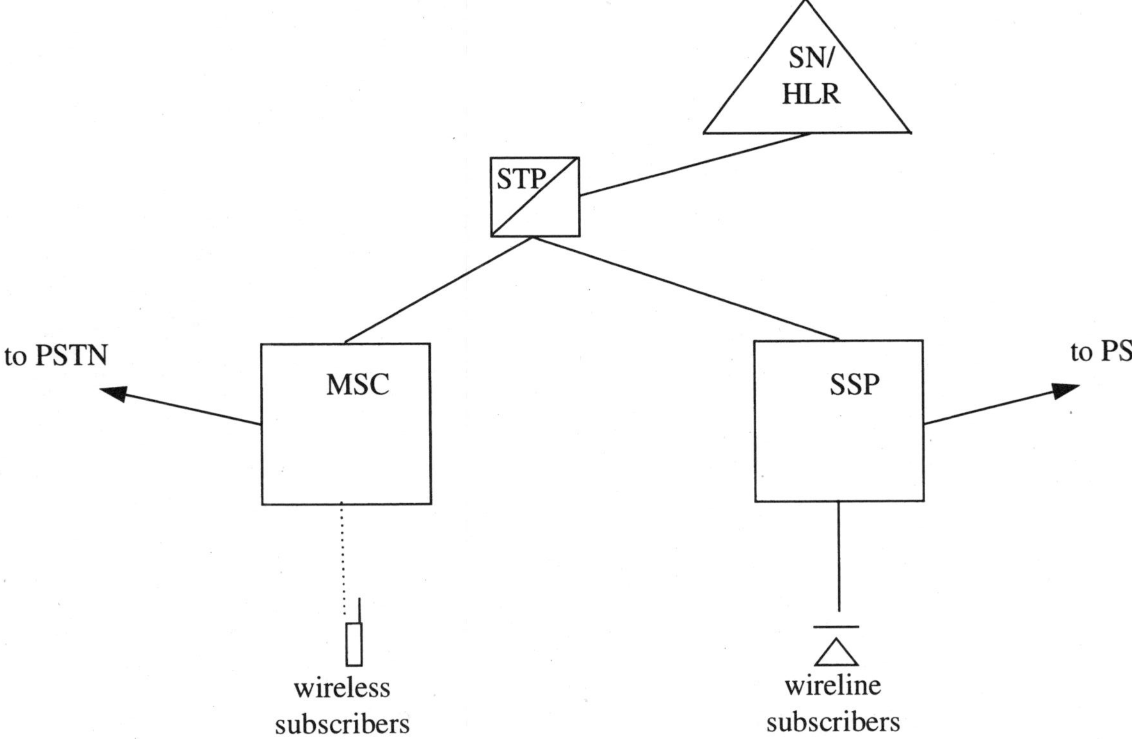

Figure 9.12 Alternative 5 without an SCP.

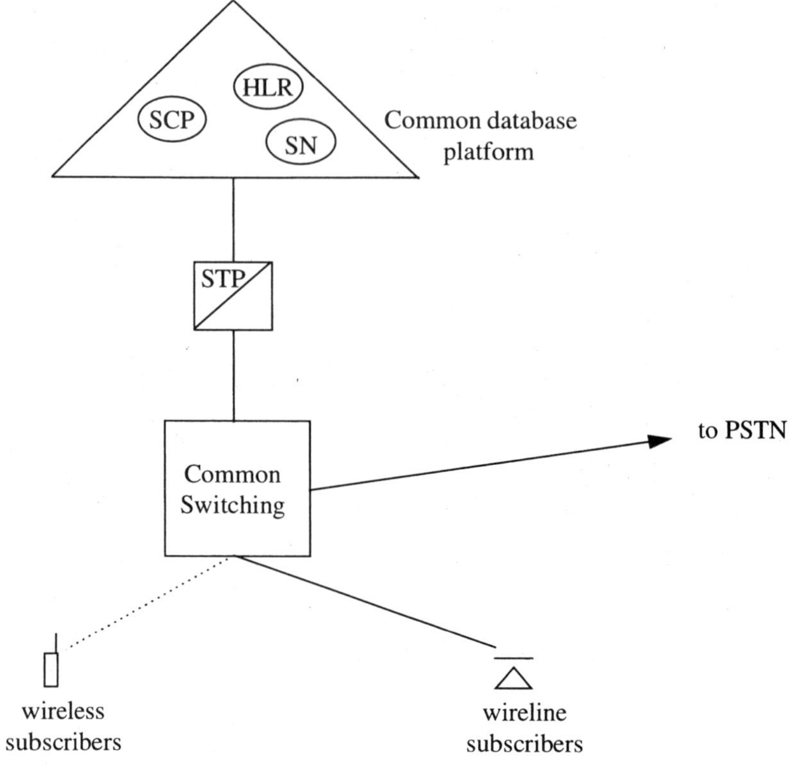

Figure 9.13 Alternative 6.

and use the same application protocols. This arrangement would allow a dynamic redistribution of applications among the database systems when changes are due.

The applications contained in the common platform do not have to be only those discussed so far. For example, in all previous alternatives a VLR was assumed to be collocated with an MSC or a common switching function. There are instances when a VLR cannot be a part of an MSC: (1) when a given system capacity does not permit a VLR function to be incorporated and (2) when a generic "C" interface is used. In either case a VLR can be contained within one of the common database platforms or provided as a separate unit. As with the rest of network elements that can be combined, capacity considerations, along with other issues, dictate the optimal location for an application. However, from the functional point of view a VLR is just another application that could be included in the common platform. The same is true of another possible application discussed previously: a

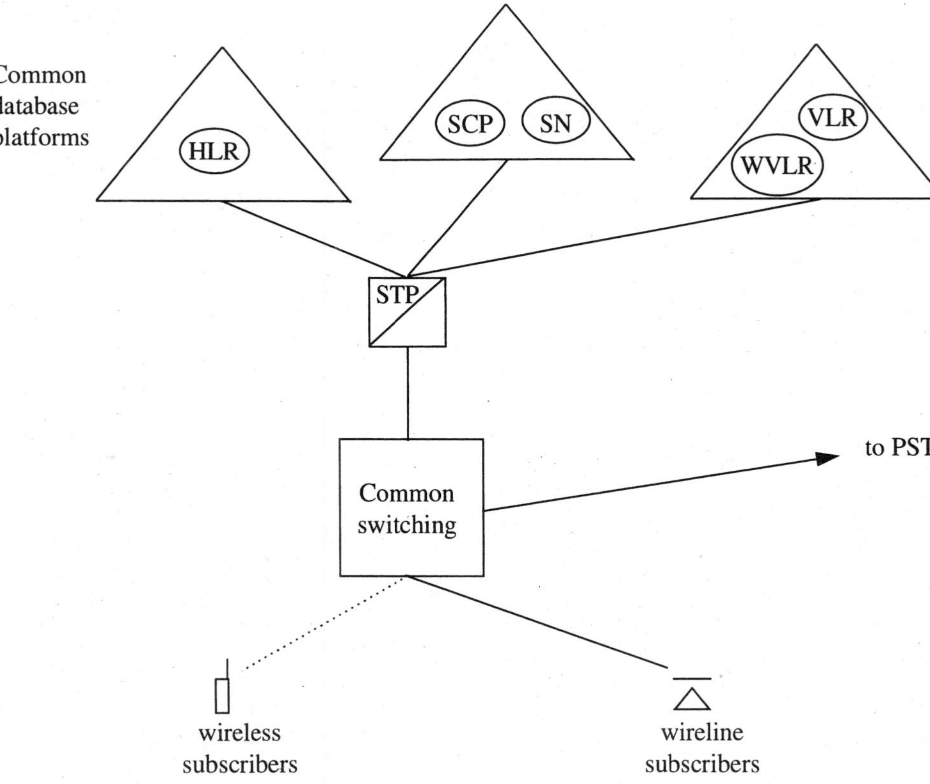

Figure 9.14 Alternative 6 with distributed common platform.

wireline VLR (WVLR). It also can be collocated with the rest of applications in the common platform.

A common platform provides many benefits to its users. It has the same set of interface, management, security, and other functions. It has the support of a common service creation environment, and it saves on training costs. A common platform allows flexible distribution of necessary applications as particular situations dictate, and it also stimulates the development and implementation of common protocols with the potential of easier migration to new capabilities and services. The final network architecture alternative is shown in Figure 9.14.

Chapter

10

GSM Issues

Up to now, all interoperability issues derived from the integration of wireline and wireless networks or different wireless networks addressed a single networking protocol that all radio technologies used in the United States. This protocol is IS-41. Whatever other differences these radio technologies have, their networking protocol is exactly the same. Similarly, in the comparison of the wireline and wireless application protocols, the same IS-41 protocol was compared to the Advanced Intelligent Network (AIN) protocol. However, there is another radio technology which is used in the United States that does not depend on IS-41; it also has other unique components built in. This technology is Global System Mobile (GSM).[1] Because the networking protocol used in GSM is considerably different from IS-41, integration of GSM and non-GSM wireless networks requires special attention. Comparison of the GSM application protocol with the AIN protocol is also required to understand the issues of integrating a GSM network with a wireline one. So, a specific look at GSM and its components is warranted.

GSM is a unique mobile radio technology. It differs from all other technologies we discussed in this book in one significant way: GSM has been defined as an end-to-end system. One batch of GSM standards contains its architecture, functional descriptions, protocol architecture, and protocol elements for all parts of the total system. This total approach created not just pieces required to support a wireless service but an end-to-end solution. In the United States several different technologies appeared, all relying on the same networking protocol, but the GSM solution "maps" or "relates" to stress the reliance

[1] See Chap. 4 for the discussion of this term.

of one piece on another. It relates one specific radio technology with a specific networking protocol, essentially eliminating the need to map the access and networking protocols.

Protocols for GSM- and IS-41-based systems perform basically the same functions, but they differ at each interface in the way these functions are accomplished. Why then do we address these differences from the networking point of view? One reason is that it is the key to network procedures. Another reason is that air interfaces are different for different technologies anyway, but in the case of GSM, because it is an end-to-end system, the access protocol is closely tied to the GSM networking protocol.

10.1 GSM Architecture

We begin the discussion of the GSM architecture with the reference model (see Fig. 10.1). It has been developed specifically for the DCS 1900 standard, but it shows the same elements and the same relationships as in the European standard. In the GSM reference model we see elements familiar to us from the discussion of Personal Communications Services (PCS) and cellular technologies with some different names. Some elements do not require explanations, such as a visitor location register (VLR), home location register (HLR), authentication center (AC), and equipment identity register (EIR). These perform the same functions as identically named elements in the cellular reference model (see Fig. 4.4). Others require some explanation:

- *PCSC.* Personal Communications Switching Center; this is a switching center comparable to the MSC.
- *BSC.* Base Station Controller; this is the same element as the BSC shown in Fig. 4.2. It controls one or more base stations.
- *BTS.* Base Transceiver Station; this is the same element as the base station (BS) shown in Figs. 4.2 and 4.4. It represents a cell.
- *RS.* Radio System; this is a combination of the BTS and BSC. This combined element is called the Base Station Subsystem (BSS) in the European GSM standard.
- *PS.* Personal Station; this element is called the MS in the cellular model (see Fig. 4.4).

To maintain some level of comfort with terminology, for the rest of the GSM-related discussions we will use the term Mobile Switching Center (MSC) (as it is called in the original GSM specifications) in place of PCSC and the term Mobile Station (MS) in place of PS.

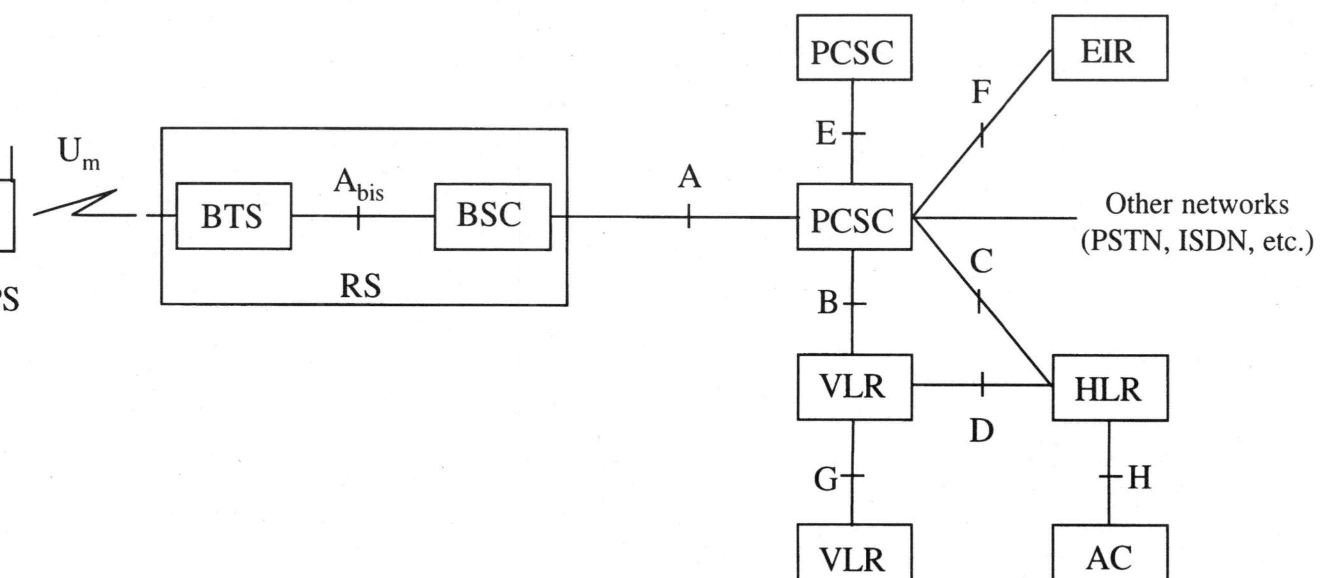

Figure 10.1 DCS 1900 reference model.

A functional configuration of a GSM network is shown in Fig. 10.2, which presents a BSC controlling several BTSs and connected to an MSC. So far everything is as we have seen with the cellular networks. One network element is new; this is a Gateway MSC (GMSC). The GMSC aids in the process of routing a call to a GSM subscriber. The GMSC connects a GSM network to a PSTN and accesses an HLR to find the location of the called party and provide the routing information to a PSTN. The GMSC function can be located in a Public Switched Telephone Network (PSTN) or in a GSM network. The physical placement of a GMSC in the GSM network can be either as an independent switch or as part of an MSC. Practical GSM implementations usually include a GMSC as part of an MSC. A VLR is also usually included in the same installation. The resulting typical GSM network architecture is shown in Fig. 10.3.[2] In it a GSM switching center looks (functionally) the same as the similar cellular or PCS switching center we discussed before. It is connected to a Signaling System #7 (SS7) network for the purposes of signaling; access to HLRs and other network elements is through the SS7 network.

10.1.1 Protocol architecture

Development of the GSM protocols has been based on the ISDN protocols for several reasons. First, GSM is a digital system and requires digital protocols. Second, Integrated Services Digital Network (ISDN) protocols had been largely developed by the time GSM development started. These protocols are Digital Subscriber System 1 (DSS1) for network access and SS7 for networking. Both are prominent in GSM. And third, since GSM networks would have to exist next to ISDNs, and probably interwork with them, it is beneficial to use protocols that foster the interworking between GSM and ISDN.

All GSM functions are grouped into one of the following categories:

- Radio resource management (RR)
- Mobility management (MM)
- Communication management (CM)

RR is responsible for the radio aspects of communications. Management of radio channels and radio equipment (e.g., BTSs) is accomplished by their functions. Control of radio resources is mostly a

[2] The authentication center (AC) and EIR are not shown for simplicity. Their functionality and connectivity is as we discussed in previous chapters.

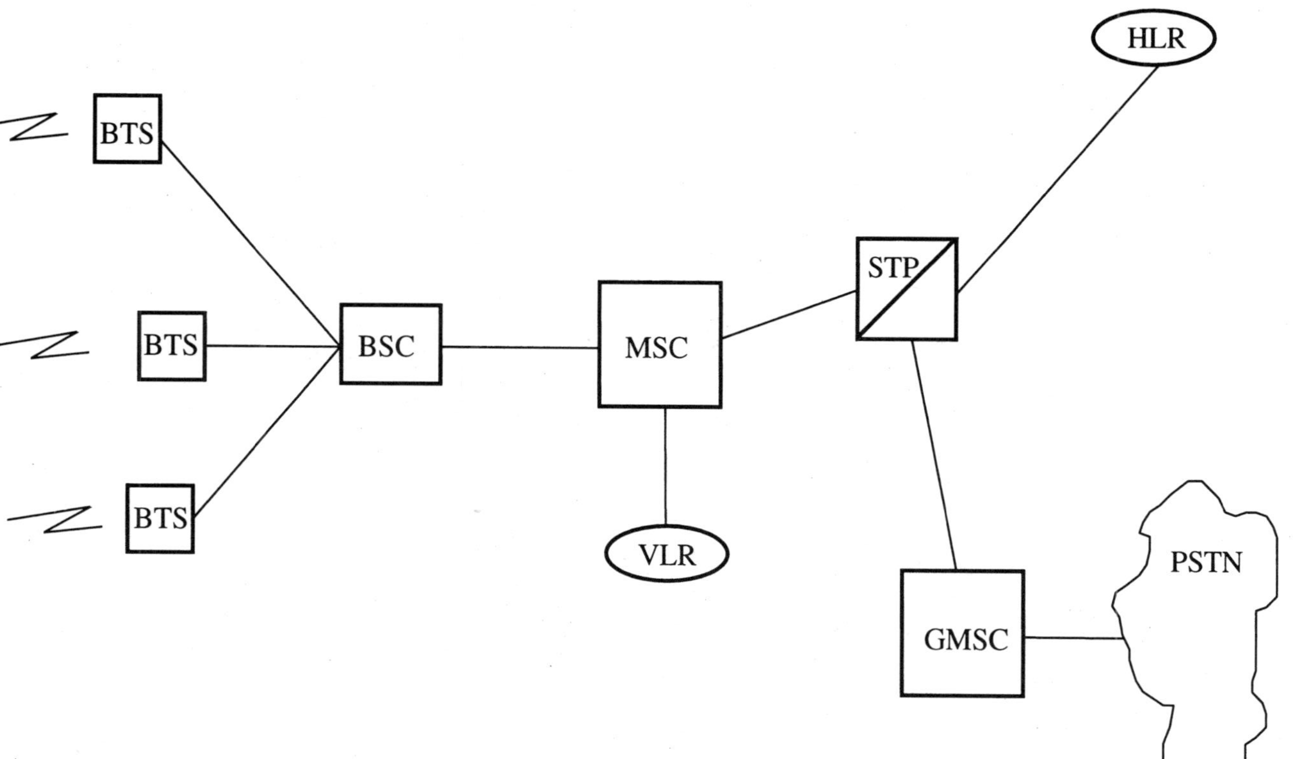

Figure 10.2 Functional configuration of a GSM network.

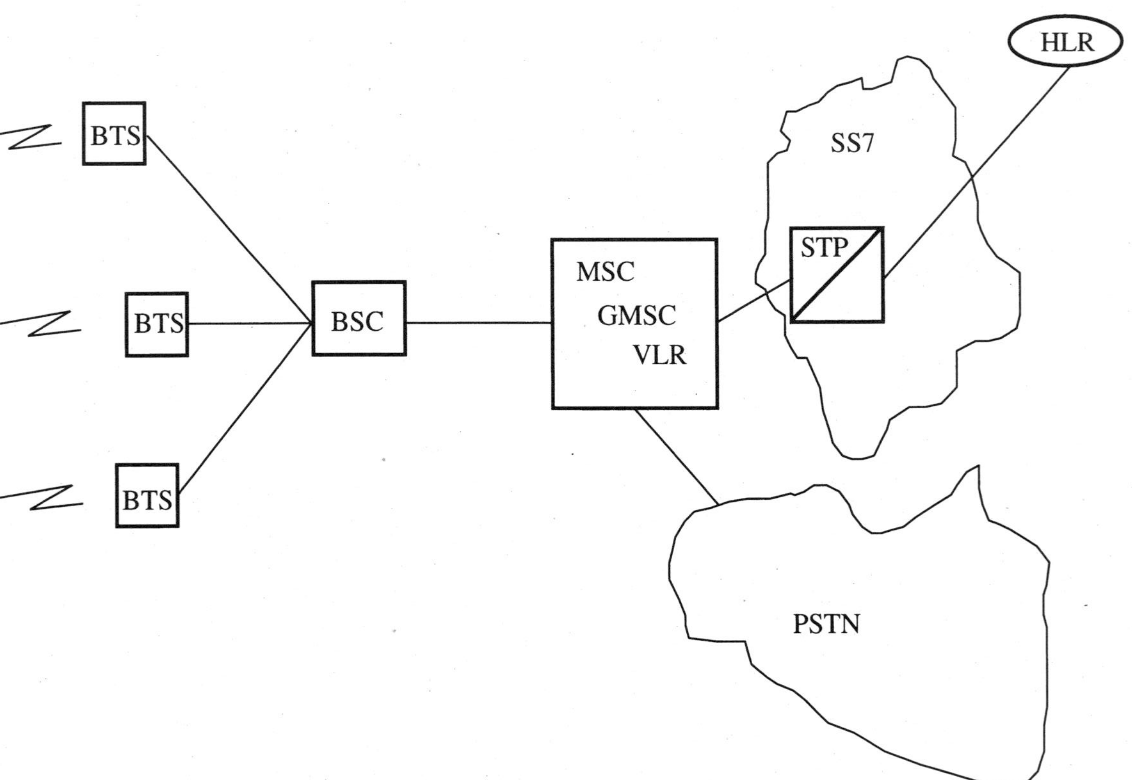

Figure 10.3 Typical GSM network architecture.

local matter, and they are managed by an MSC and BSSs. MM deals with the mobile aspects of subscribers, such as registration, roaming, and other such functions. MM involves an MSC and supporting databases (VLR, HLR). CM is responsible for establishing, clearing, and otherwise manipulating (e.g., for supplementary services) calls. It involves an MSC, VLR, HLR, and PSTN.

Figure 10.4 shows the protocol stacks that support each of these categories. On the access side Direct Transfer Application Part (DTAP), RR, and BSS Management Part (BSSMAP) are application protocols that are used for various GSM functions. These application protocols use different underlying protocols as transports. MM and CM are supported by the DTAP protocol. This is an application protocol that supports communications between an MS and an MSC. Over the air interface DTAP relies on the Link Access Protocol on the D-Channel ($LAPD_m$) protocol—an adaptation of the LAPD protocol for the air interface—for data transfer. The LAPD protocol is used to transfer DTAP messages between a BTS and a BSC, while over the A-interface Signaling Connection Control Part/Message Transfer Part (SCCP/MTP) protocols are used.

For RR a protocol is split into two parts. Over the air a specific GSM radio protocol (we call it RR) is used, supported by LAPDm and LAPD protocols. On the A-interface the BSSMAP protocol is used, supported by an SCCP/MTP. BSSMAP and RR are application protocols performing radio-related functions.

Between a BSS and an MSC two application protocols, BSSMAP and DTAP, are used directly over SCCP. A distribution function (DF) is therefore deployed to differentiate traffic on the A-interface between these two protocols.[3]

The networking protocol supports MM functions and is called Mobile Application Part (MAP). It has been developed as an extension to the suite of SS7 protocols and uses SS7 transport for the exchange of information between various network entities.

An important point about the GSM call control protocol is that it works just like it does in ISDN. That is, an application protocol based on DSS1 acts between an MS and an MSC, which means that a terminal (in this case an MS) is controlled from the switch (an MSC). The intermediate nodes (BTS, BSC) are there to relay call control information. Beyond the MSC, in the network, the SS7-based call control application protocol is used. Figure 10.5 shows this concept.

---

[3] The DF is a substitute for some functions Transaction Capabilities Application Part (TCAP) can provide. The DF is a lot simpler and sufficient for the job. It does make for a nonstandard SS7 implementation.

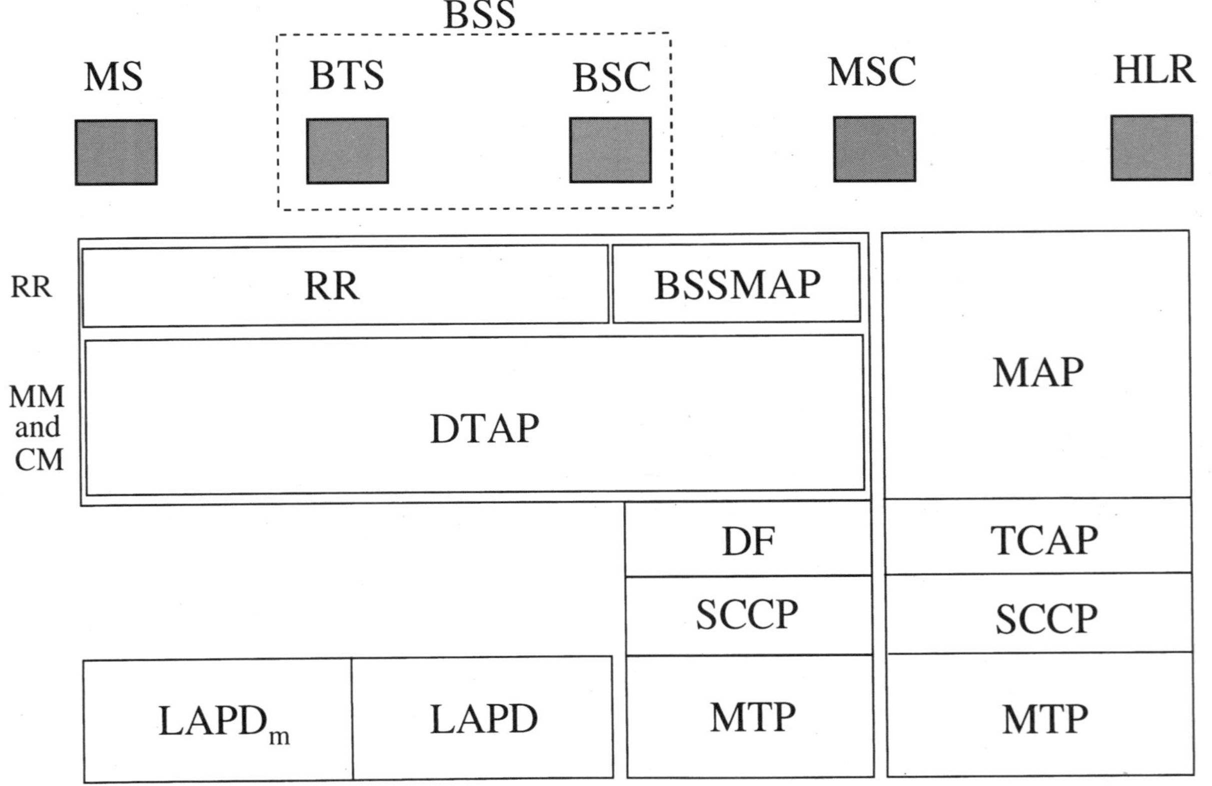

Figure 10.4 GSM protocol stacks.

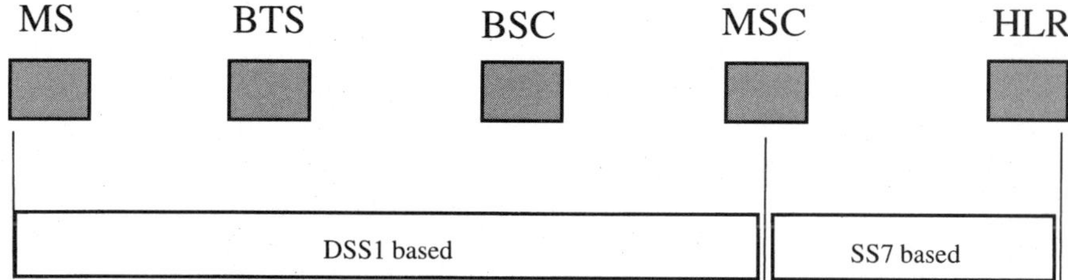

Figure 10.5 GSM call-related protocol model.

10.2 Services

We already mentioned most GSM services in the list of cellular services. Obviously, voice services are provided together with mobility services such as roaming and handoffs. Various data services are offered at different speeds up to 9.6 kbps (using current transmission techniques). The number of supplementary services is quite large; we will list just some of them here:

- Voice privacy
- Call forwarding (different flavors are available)
- Call waiting
- Multiparty calling
- Calling line identification presentation
- Call hold
- Preferred language
- Advice of charging
- Call barring

One service has been specifically designed for GSM—short message service (SMS).[4] It consists of the delivery of a small amount of data to a handset for immediate display. A short message is several dozen characters long.

---

[4] Although SMS was created as a GSM service, currently it is also supported by IS-41 Rev. C.

10.3 GSM MAP

The GSM MAP protocol is an application layer protocol, and it consists of a number of messages. These messages are transported using TCAP and the rest of the SS7 protocol stack including SCCP. In principle the GSM MAP is built in a fashion similar to other application protocols, such as IS-41 or AIN. The differences in content and meaning of the messages exist for two reasons: (1) GSM procedures are different and (2) GSM and IS-41 protocols were designed by different people. We will discuss some specific examples of procedural differences in the following sections, but right now we will compare GSM MAP and IS-41 protocols.

Let's examine two messages performing the same functions in DCS 1900 MAP and IS-41. The function is to request routing information; the messages are sent from an MSC to an HLR. The messages are Map_Send_Routing_Information in GSM MAP (presented in tabular form as Table 10.1 and 10.2) and LocationRequest in IS-41 (see Tables 10.3 and 10.4).

Some of the parameters in the DCS 1900 message are presented here for comparison with the IS-41 parameters:

Table 10.1 Map_Send_Routing_Information Request

Parameter	Type*
PSISDN	M
Number of forwarding	O
Network signal information	O

*M = Mandatory; O = optional.

Table 10.2 Map_Send_Routing_Information Response

Parameter	Type*
IMSI	O
PSRN	O
Forwarding data	O
PrefCarrierIDList	O
User error	O

*O = optional.

Table 10.3 LocationRequest Request Portion

Parameter	Type*
BillingID (originating)	M
Digits (dialed)	M
MSCID (originating)	M
SystemMyTypeCode (originating)	M
CallingPartyNumberDigits1	O
CallingPartyNumberDigits2	O
CallingPartySubaddress	O
MSCIdentificationNumber	O
PC_SSN (originating)	O
RedirectingNumberDigits	O
RedirectingSubaddress	O
SenderIdentificationNumber	O
TerminationAccessType	O
TransactionCapability	O

*M = Mandatory; O = optional.

- *PSISDN.* Personal Station ISDN Number. Directory number (DN) assigned to a called subscriber called Mobile Station ISDN Number (MSISDN) in the European Telecommunications Standards Institute (ETSI) GSM standard
- *Number of forwarding.* Forwarding number
- *IMSI.* International Mobile Subscriber Identification; subscriber identifier
- *PSRN.* Routing DN
- *PrefCarrierIDList.* List of preferred network carriers

Although the two presented GSM messages are actually defined as one message in the DCS 1900 standard, the meaning is the same: The standard message consists of the request and response portions. We just split them into two messages. They are shown this way because

Table 10.4 LocationRequest Response Portion

Parameter	Type*
ElectronicSerialNumber	M
MobileIdentificationNumber	M
MSCID (serving MSC)	M
AccessDeniedReason	O
AnnouncementList	O
CallingPartyNumberString1	O
CallingPartyNumberString2	O
Digits (carrier)	O
Digits (destination)	O
DMH_AccountCodeDigits	O
DMH_AlternateBillingDigits	O
DMH_BillingDigits	O
DMH_RedirectionIndicator	O
GroupInformation	O
MobileDirectoryNumber	O
NoAnswerTime	O
OneTimeFeatureIndicator	O
PC_SSN (serving MSC or VLR)	O
RedirectingNumberDigits	O
RedirectingNumberString	O
RedirectingSubaddress	O
RoutingDigits	O
TerminationList	O
TerminationTriggers	O

*M = Mandatory; O = optional.

IS-41 messages are defined using this representation. The meaning of these messages (or rather two portions of the same message) is that the request is sent with some parameters, and the response to this request contains the same or other parameters or a combination of them.

Some of the IS-41 parameters are:

- *BillingID.* Billing identity
- *Digits (dialed).* Dialed digits
- *MSCID (originating).* Switch identification
- *ElectronicSerialNumber.* Handset identification
- *MobileIdentificationNumber.* Identification of the subscriber
- *AccessDeniedReason.* Reason for the denial

A comparison of parameters in DCS 1900 and IS-41 messages shows that they have very little in common. DCS 1900 messages that request and return routing information are short, while their IS-41 counterparts contain many parameters. Although most parameters in these IS-41 messages are optional, when present they provide additional information which can be used to support more services.

This comparison shows how different the two networking protocols are. Interworking between them would require significant effort either in changing existing equipment or adding special interworking functions.

Although currently GSM MAP does not contain IN-specific functionality, it would be safe to assume that when the GSM community decides to model and create protocol standards to support IN functionality [something like the Wireless Intelligent Network (WIN) for GSM], the MAP will be the basis for such development. Only when this occurs will we be able to discuss possible differences between this GSM protocol and IS-41 WIN capabilities or the AIN protocol.

10.4 Identification

Four identifiers are used in GSM to address a subscriber and the associated handset. These are:

- IMSI (International Mobile Subscriber Identification)
- TMSI (Temporary Mobile Subscriber Identification)
- MSISDN (Mobile Station ISDN *Number)*
- IMEI (International Mobile Station Equipment Identity)

IMSI and TMSI are used to identify a subscriber, not a handset. MSISDN is a dialable number assigned to a given subscriber. IMEI identifies a particular handset.

There are several differences between these and the identifiers used in IS-41-related radio systems. First, in non-GSM systems the same number is used to identify and address a subscriber—the functions performed in GSM by IMSI and MSISDN. Second, non-GSM systems do not currently use TMSI, which is a temporary identifier assigned to a subscriber for a relatively short duration. TMSI is used over the air most of the time to preserve the user's privacy. There are additions

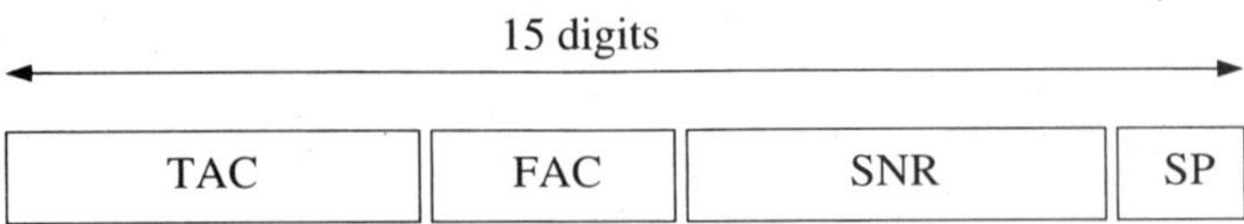

TAC - Type Code Approval; 6 digits
FAC - Final Assembly Code; 2 digits
SNR - Serial Number; 6 digits
SP - Spare; 1 digit

Figure 10.6 EMEI breakdown.

to the U.S. air interface standards to accommodate both the use of temporary identifiers and subscriber identifiers independent of a DN (especially in CDMA). But at this time no such capabilities exist in IS-41.

There is also another difference. The IMEI identifies a handset in GSM and the Electronic Serial Number (ESN) performs the function in IS-41. However, their formats are different. The IMEI consists of 15 digits, as shown in Fig. 10.6. The ESN consists of 32 bits, the first 8 of which identify a manufacturer and the rest identify a serial number. Obviously, these two identifiers are incompatible.

10.5 Procedures

GSM performs the same basic functions as other radio technologies. It supports call origination and termination, handoffs, authentication, registration, paging, and so on. If we analyze GSM procedures, we find that several of them are substantially different. Let's examine them.

10.5.1 Registration

When a GSM subscriber is registering in a new location, the location update procedure is invoked. Figure 10.7 shows this procedure when a new VLR serves the new location. The following takes place:

1. An MS requests registration; it sends TMSI as a subscriber identifier.
2. A BSS transfers the TMSI to the MSC with the registration request.
3. The MSC transfers the TMSI to the new VLR.
4. The old VLR is queried.
5. The old VLR sends the subscriber's IMSI to the new VLR.
6. The new VLR sends registration information to the HLR.
7. The HLR responds with the registration.
8. The new VLR registers the subscriber and informs the MSC.
9. The BSS is updated.

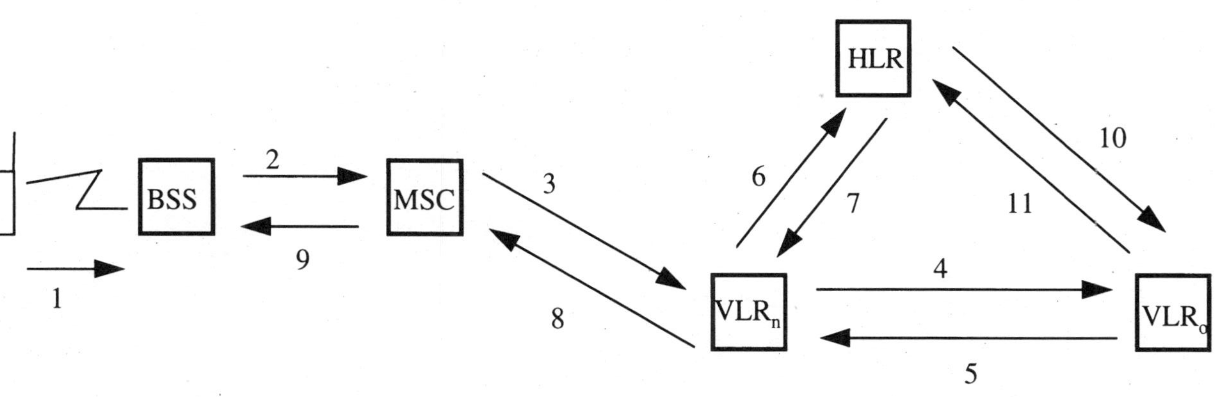

VLR$_n$ - new VLR

VLR$_o$ - old VLR

Figure 10.7 GSM location update procedure.

10\11. Deregistration at the old VLR is performed.

This procedure is very similar to registration procedures performed by radio technologies using IS-41 as the networking protocol. The difference lies in the VLR-to-VLR communications in GSM. It is needed because IMSI is required for the registration to take place. However, an MS sends a temporary value, TMSI, over the air, leaving a new VLR without the IMSI. This identifier, in relation to a particular TMSI, is stored at the old VLR. Only after receiving the IMSI from the old VLR can the registration procedure be completed. This unique registration process makes it completely incompatible with IS-41-based systems.

10.5.2 Handoffs

Handoffs in GSM are controlled by the network. As in non-GSM systems, handoffs can be intra- and inter-BSS, and intra- and inter-MSC. The principles of any type of handoff are the same, so, for simplicity, we will consider only the inter-BSS and intra-MSC handoff. Figure 10.8 shows a simplified GSM handoff procedure. Here are the actions:

O. An MS communicates with the current BSS (BSSo). While this communication goes on, the BSS monitors its quality by performing measurements. It also receives reports from the MS on measurements performed there. When certain criteria are met, the BSS decides that a handoff should be performed.

1. The old BSS (current BSS) informs the MSC of the need for a handoff.

2. The MSC requests a handoff to a target BSS (BSSn).

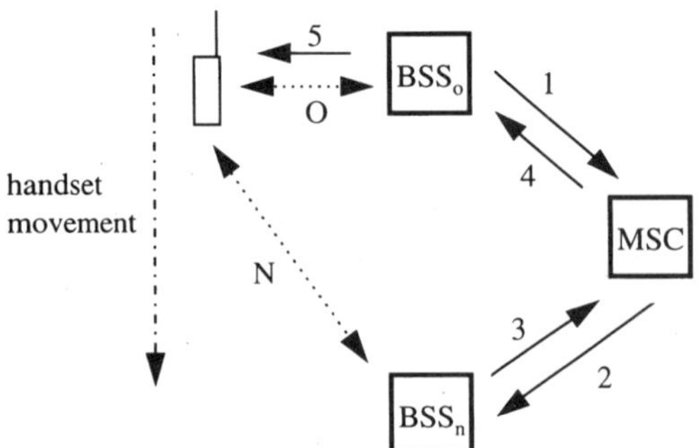

Figure 10.8 Simplified intra-MSC handover in GSM.

3. The target BSS (new BSS) grants the request.
4. The MSC informs the old BSS that the handoff is granted.
5. The old BSS informs the MS of a new channel to be used.
N. The MS switches its communication channel to the new BSS.

At this point the MS is controlled by the new BSS. It begins measurements of the MS communications. Actually, a formal handoff procedure still continues after the initial contact between the MS and the new BSS has been established. At this point the new BSS reports this activity to the MSC, which then requests the old BSS to clear the previously used channel. But the main point of this presentation is that in GSM the handoffs are initiated and performed in a source-to-target fashion under the control of a BSS.

In IS-41-based systems the philosophy of handoffs is very similar. Hence, there should not be big problems in making seamless handoffs. However, this applies only to the networking portion of handoffs when MSC-to-MSC handoff is required. There is also a difference between high-power radio systems (e.g., GSM, CDMA) and low-power systems, such as Personal Access Communication System (PACS). In high-power systems, handoffs are usually accomplished source-to-target, but in PACS (and other low-power radio systems) handoffs are usually accomplished target-to-source.

The GSM example in Fig. 10.8 presents a handoff initiated by the current system (source). In target-to-source systems, a handoff is initiated by the new (target) systems. Figure 10.9 shows such an example. Here is what happens:

O. The handset communicates with the old BS (current BS) and monitors quality of communications.
1. When a signal from a new BS meets specific criteria, the handset requests a handoff to be performed.
2. The new BS forwards this request to the MSC.
3. The MSC informs the old BS of the handoff request.
4. The old BS sends appropriate information to the MSC.
5. The MSC forwards the necessary information to the new BS.
6. The handset is informed that the handoff can take place.
N. The handset uses a communication channel to the new BS.

The conclusion of this discussion is that if a seamless network is to be created by combining technologies with different handoff philoso-

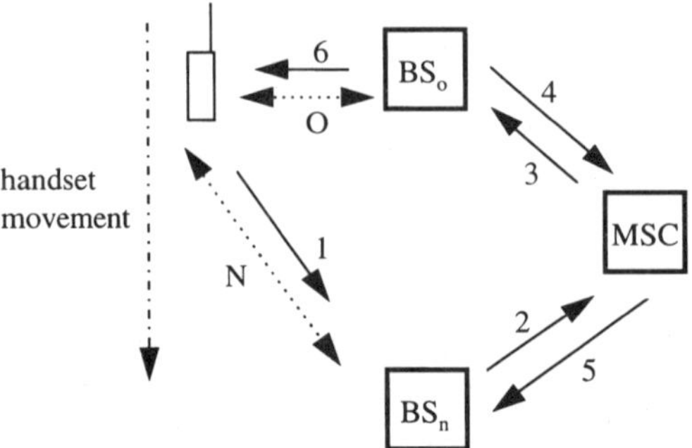

Figure 10.9 Simplified intra-MSC handover in PACS.

phies, the differences in handoffs have to be carefully analyzed to make this process seamless. Obviously, such an analysis would be required when it was determined that a service of handing off calls from one technology to another is desirable, as in the service-rich multitier arrangement discussed in Chap. 11.

10.5.3 Authentication

Authentication is the process of confirming a subscriber's privileges. It is performed to minimize the possibility of fraud in wireless networks. While the idea and the need for authentication are the same in GSM- and IS-41-based systems, the implementations are different.

The basic concept of authentication is based on the private key mechanism, which is a value used in specific authentication algorithms and assigned to a particular subscriber (private). A generic private key procedure consists of one side encoding some value using a private key and the other side decoding the resultant information with the same private key and using a reverse algorithm. If the calculated value equals the original, the authentication process is successful (see Fig. 10.10). Although both IS-41 and GSM MAP use this generic concept, there are differences in the authentication algorithms and the sequence of events that takes place.

In GSM a private key is called K_i and is stored at the AC and in the subscriber identity module, or SIM (see the following section of this chapter) of a particular subscriber. This key is never transmitted. The authentication algorithm used in GSM is A3. The authentication mechanism is implemented by creating so-called triplets of values in the AC and storing them in the VLR for each subscriber. The triplets consist of RAND (random value), SRES (signed response), and K_c (ciphering key used for link ciphering). Every time an authentication of a subscriber is to be performed, one of the triplets is used in the

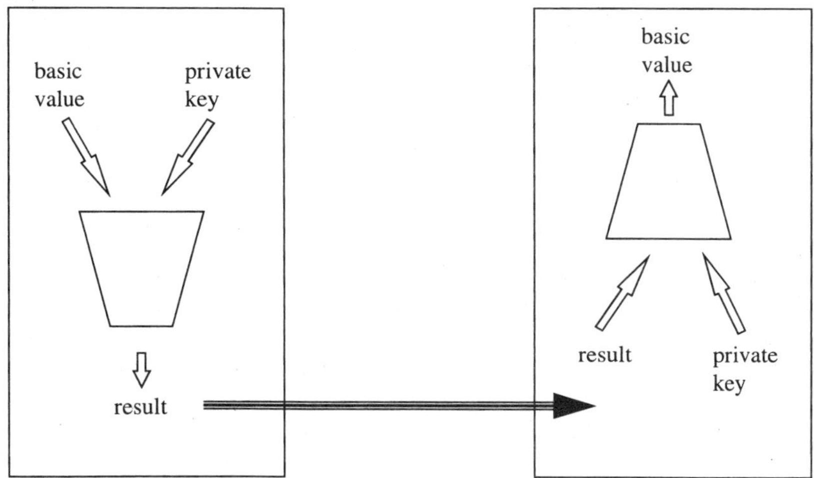

Figure 10.10 Conceptual private key authentication procedure.

process. When all triplets have been used, a new batch of them is received from the AC.

In IS-41 a private key is called an A-key. This key is used for the production of the intermediate values called *shared secret data* (SSD), which is used in the authentication procedure. The authentication algorithm used in IS-41 systems is Cellular Authentication and Voice Encryption (CAVE).

Given different authentication algorithms and different procedures, it is clear that incompatibility between GSM MAP and IS-41 in this area is significant. It would present a barrier for seamlessness.

10.6 SIM

Another unique feature of GSM is a Subscriber Identity Module (SIM). SIM is a removable component of a GSM handset. It is either credit card-like and easily removable or a smaller-size chip which is not difficult to remove but would require some actions to do so. SIM is like an additional computer in a handset that allows storage of information, and, potentially, decision-making regarding services. The type of information stored in SIMs depends on the services they support, but the basic categories of information always present in a SIM are (1) subscriber identity and (2) authentication. An example of subscriber identity data is IMSI; an example of authentication is a K_i key used in the authentication algorithm.

SIM acts as a key to opening up a service to a subscriber. Without a SIM installed in a handset no service can be provided (except an emergency call service). This feature provides extra security protection and, at the same time, allows a subscriber to use different handsets while traveling or under other circumstances, such as the need for handset repair. Traveling with a SIM is easier than traveling with a handset, so the SIM concept helps the subscribers who move about.

The SIM can be a repository of service-related information. It can store some abbreviated numbers, accept short messages, or even maintain a database associating certain DNs with names and locations of their owners. These kinds of services are the future for SIMs, and they will definitely grow in complexity.

The SIM concept is so promising from the services point of view that it is being considered as an addition to IS-41 based technologies in the United States. Introduction of SIMs in the United States will probably coincide with the extension of subscribers' identifiers to encompass IMSI and TMSI. However, at the present time the handsets developed for the IS-41-based radio technologies do not contain SIMs. This probably does not present a problem from the integration point of view; an important integration issue is not where the data are stored (in the removable or fixed part of a handset) but what kind of data are used.

The only SIM-related integration issue is the type of services supported by GSM handsets equipped with SIMs and other handsets. This may become a driving force for SIMs to be introduced in the United States.

10.7 Concluding Remarks

There are considerable differences between GSM and other radio systems targeted for U.S. deployment that would make the task of integrating them quite complex. Of course, the degree of desired integration is important in order to understand what is to be done. If the goal is a complete seamlessness of GSM and non-GSM networks, the task is difficult and requires modifications to existing systems. If sharing of the databases is the goal, the task is simpler, but not simple.

In addition to the issues we already covered, two aspects of GSM development and deployment should be mentioned. The first is the underlying SS7 transport mechanism which is the carrier of both GSM MAP and IS-41. Since GSM MAP is a European standard and IS-41 is an American one, the SS7 they use is, respectively, European (ITU-T) and American (ANSI). We already covered differences between the two SS7 standards, and they all apply to GSM. In addition, there are differences in the digital carrier systems that have to be alleviated if GSM is to be used in the United States (obviously, if American-developed technology is to interwork with GSM networks in countries where ITU-T SS7 standards govern SS7 implementations, the U.S. technology would have to adapt to the foreign environment). For example, in the United States a T1 carrier supports 1544-kbps transmission over 24 channels, while a European E1 carrier supports 2048-kbps transmission over 32 channels.

The second aspect of GSM we need to mention here is related to the development of standards in the United States. The MMAP protocol that T1S1 is developing to support a generic network access contains elements in support of various radio systems, including GSM (DCS 1900). It is difficult to predict how much impact the MMAP will have on the process of integrating GSM and non-GSM networks, but it is encouraging that such efforts exist.

Chapter

11

Network Access Issues

Network access plays a pivotal role in providing services. The capabilities of network access interfaces largely determine what kind of services can be offered to subscribers. As an example, if the network is capable of producing information for a customer, but an access interface is not capable of using this information or delivering it to a customer, no service relying on this information can be supported. Attempts to integrate various networks and make network services seamless include various ideas for network access interfaces.

11.1 Generic Wireless Network Access

When we discussed various wireless technologies, it was obvious that using a variety of these technologies in the same network would make this network too complex. If, for any reason, the same network is required to handle different wireless access technologies, a technique to simplify the situation would help. If such a situation seems unlikely, look at the existing public networks with large investments in the infrastructure and the need to use this available equipment to support emerging services.

11.1.1 Generic access architecture

The idea of a generic wireless access to wireline networks has been developed by Bellcore and is known in the industry as the *generic C interface.* This name includes an interface reference because it is based on the Personal Communications Services (PCS) reference configuration defined by T1P1 (see Fig. 4.10). In this reference configuration a "C" interface is basically an interface between a network and radio equipment. The translation of this aspect of the PCS reference

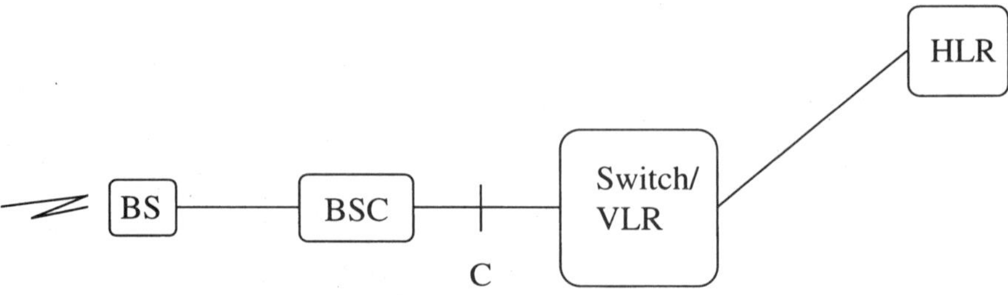

Figure 11.1 Typical physical network configuration.

configuration into a typical physical network topology is shown in Fig.11.1.

The switch in this configuration is expected to work with a particular base station controller (BSC) or a base station (BS) which is designed for a particular air interface, [e.g., Code Division Multiple Access (CDMA), Global System Mobile (GSM), or Personal Access Communications System (PACS)]. If at the "C" interface (between a BSC and a switch) a protocol can be defined that handles multiple air interfaces, such a protocol would be generic, and an interface that supports this protocol is called a *generic interface.*

The goal of the generic "C" interface is to support air interfaces that may be used in the United States. The most probable users of this concept are public networks. The switches deployed in the public networks are service switching points (SSPs), not mobile switching centers (MSCs) that usually have a visitor location register (VLR) integrated within a switch. The target architecture of the network access using a

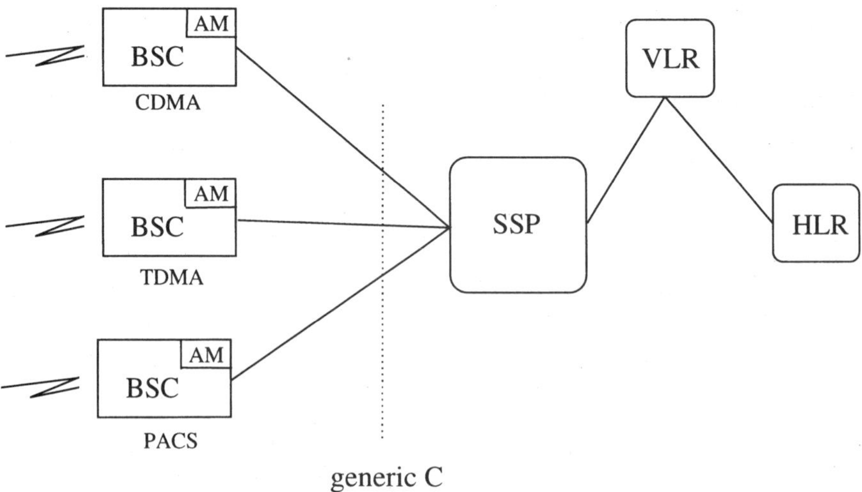

Figure 11.2 Generic "C" interface.

generic interface is shown in Fig.11.2. It depicts several different radio systems communicating with the network over the same interface.

The functions of an access manager (AM) are specific to a particular radio technology. Furthermore, depending on the implementation, the functional split between a VLR and an AM can be different. For example, in PACS a call-related identifier, called alert_ID, is managed by the PACS AM. The physical location of an AM is not specified in the reference configuration. It can be collocated with a BSC or not. Since AMs are radio-specific, placing them at the radio equipment side opens the possibility of hiding differences between different radio technologies within these units and using a common interface beyond the radio to the network. This is the main premise of the generic interface concept.

Two types of activities take place over the "C" interface: call control and mobility management. Although the division is not formally precise, it does help to classify functions and communications interfaces. Call control activities are those related to functions in the switch, such as alerting, routing, and switching. Mobility management activities are related to registration, roaming, and authentication and generally do not involve the switch, but rather a VLR. Relating these two types of activities to functional groups produces the following general communications needs: BSCs communicate with SSPs for call control purposes, while AMs communicate with VLRs for mobility purposes.

An attempt at standardizing this interface has been made with some success. TR 46 is ready to produce an interim standard (IS-653)

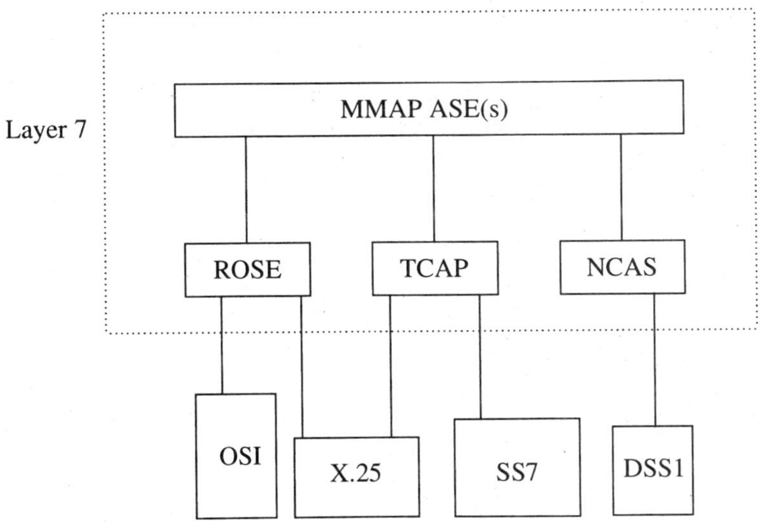

Figure 11.3 MMAP protocol architecture.

for an Integrated Services Digital Network (ISDN)-based A-interface (remember, the A-interface in TR 46 is functionally the same as the C-interface in T1P1). This standard separates functions into call control and mobility management. ISDN call control is used to support call processing needs, while Mobility Management Application Protocol (MMAP) was developed by T1S1 for mobility management purposes. The first phase of the MMAP standard (T1.651) specifically covers AM-to-VLR mobility management functions for the following technologies: CDMA (IS-95 based), DCS 1900, PACS, PCS 2000 (composite CDMA/TDMA), and Time Division Multiple Access, or TDMA (IS-54 based). This is a truly integrated part of the access interface that can be extremely valuable in the process of making networks seamless since it hides technology variations from the network as much as can be expected.

MMAP is developed as an application layer protocol independent of the underlying transport mechanisms. The MMAP application service element (ASE) can use Transaction Capabilities Application Part, Remote Operations Service Element, or non-call-associated signaling (TCAP, ROSE, or NCAS) which in turn can use the services of either X.25, DSS1,[1] or SS7 (see Fig. 11.3). While X.25 and SS7 are stack of protocols developed as transport systems (and we already discussed them), NCAS is a little different and requires some explanations.

NCAS is the type of signaling which has been developed for activities not related to direct call control. It is needed for the different requirements that call control and non-call-control applications have. TCAP is another example of a non-call-control protocol. But, while TCAP was developed for transaction processing, NCAS is a method of transporting information using a protocol which has been developed specifically for call control. DSS1 is a protocol defined for ISDN access; it is primarily a call-control protocol, but it has a mechanism by which any information can be encapsulated within a message and passed from one node to another. This is what NCAS does, and such a capability opens a door for another function that a switch can perform in the generic access environment.

Figure 11.4 shows how MMAP uses the DSS1 protocol between an AM and a switch, and uses Advanced Intelligent Network (AIN) messaging between a switch and a VLR (here, a VLR can be considered to be in the service control point, or SCP). Essentially, a switch is used

---

[1] Digital Subscriber System 1 (DSS1) is the ISDN network access protocol, which is why the author modified the figure of the MMAP protocol architecture contained in the MMAP standard (T1.651). That figure shows ISDN as the transport mechanism for NCAS, but ISDN is not a protocol but a concept and DSS1 is a specific access protocol developed for ISDN.

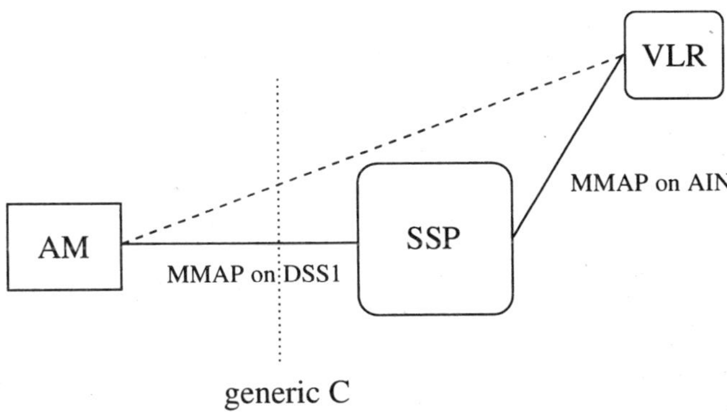

Figure 11.4 Non-call-associated signaling.

as a relay between an AM and a VLR. This is a new architectural solution which moves even further away from the traditional wireless switching environment. The MSCs used to contain not only a VLR but also a home location register (HLR). The new solution offers switching system independence from any databases and uses the already available AIN protocol for communications.

Positioning a VLR outside an SSP is not related to non-call-associated signaling. With any interface between an AM and a VLR, physically locating a VLR outside a switch is a viable solution. The advantage of this architecture is the reuse of already available AIN network elements and interfaces (see Fig. 11.5). It is an interesting architectural evolution: from an MSC with databases to a simpler

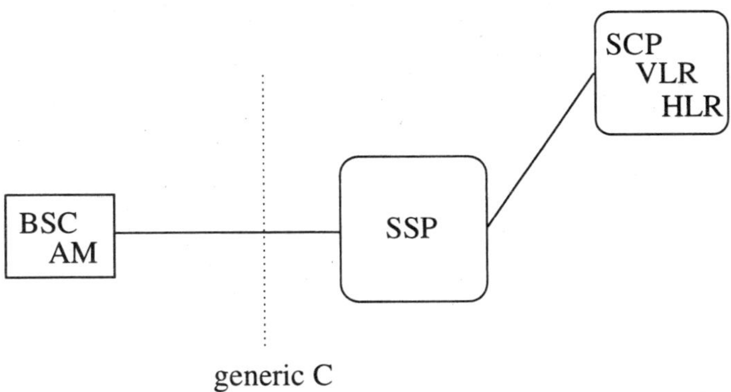

Figure 11.5 Common database.

switch with all databases collocated and separated from the switch. The potential benefits of such an architecture can be the streamlining of traffic and commonality of AIN and wireless intelligent network (WIN) services.

11.2 Wireline Registration

When a customer subscribing to a wireless service is at home next to the fixed wireline phone, he or she would like to be reached over the wireline network. It is cheaper because no air time is involved; depending on the environment, it can also provide better voice quality. At present, the only ways this can be done are to inform all potential callers that at certain times of day a home telephone number should be used, turn off the mobile handset and allow callers to guess that this person may be at home, or provide an announcement to that effect. The concept of wireline registration would create a seamless service to a customer by reporting to the wireless network information regarding the location of the customer in the fixed wireline domain (see Fig. 11.6).

The process would work as follows: A wireless service subscriber has a MIN (let's say it is 908-345-6789). When at home, this customer registers via a fixed phone with the wireless network, notifying the network of the present location. This notification has to end up in the customer's home HLR. The HLR may associate this type of request with a preassigned telephone number (which is, let's say, 908-234-5678). Another way of doing it would be to input a particular directory number (DN) while registering. When a call to number 908-345-6789 is coming in, the HLR, which is consulted to find out the customer's location, determines that the customer is currently served by a fixed phone and returns 908-234-5678 as the temporary line directory number (TLDN). When the originating switching system routes a call to this number, the fixed phone at our customer's home rings and the customer is spared more expensive air charges for a terminating call.

One functional element is missing from this picture. In order to make it work, a VLR-like function at the SSP is needed. In today's cellular networks location information is reported to an HLR by a VLR

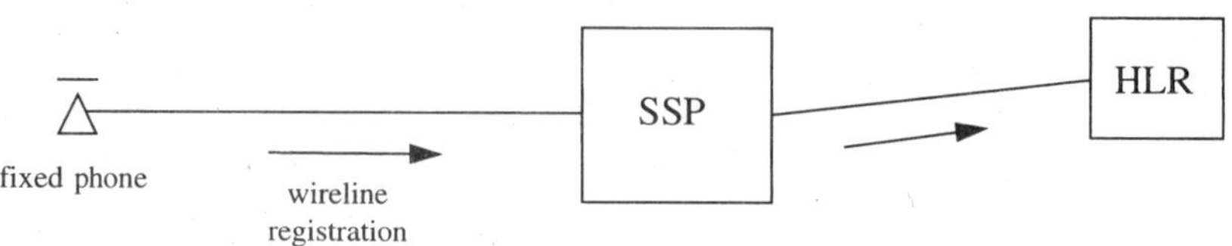

Figure 11.6 Wireline registration concept.

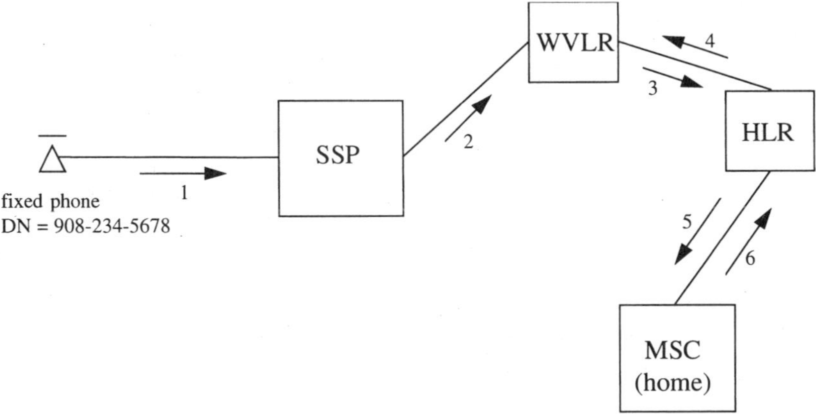

Figure 11.7 Wireline registration call flow.

(even if a VLR is physically included in an MSC). A function has to be added to an SSP to interpret the registration request from a fixed phone and to report a new location to the HLR. Let's examine the call flows in Fig. 11.7 using an element that we can call a wireline VLR, or WVLR. Let's also remove the function of associating a specific DN with the wireline registration from an HLR and allow a WVLR to report this DN as the TLDN (this would be more in line with the current activities in cellular networks).

The following describes what is occurring in the network in support of wireline registration:

1. A subscriber communicates with an SSP. This can be done in one of two ways:
   a. A subscriber manually dials a specified code or a prefix, signifying a willingness to register at this location, and an MIN that identifies the subscriber's handset.
   b. A subscriber dials a predefined number on the mobile handset and thereby communicates with a fixed base located at home; reception by the base of a code from the handset starts an automatic registration with an SSP as above. In this case a new piece of customer equipment is required—a base. This, obviously, has to be developed, but it is in the realm of possibilities.
2. An SSP passes the registration request with an MIN and a DN to a WVLR.
3. A WVLR sends a registration notification request to an HLR.
4. An HLR sends all appropriate information to a WVLR.

5. An HLR cancels the previous registration.

6. An HLR receives an acknowledgment of the registration cancellation.

Using this scheme requires the development of a wireline VLR, but, once developed, this arrangement supports a seamless service delivery to a customer. The WVLR, since it will have all the necessary profile information, can support other services for a subscriber as well. Access to a WVLR can be implemented either directly or through the SSP. In the case of direct access, a subscriber would dial an access number that would set up a connection with a WVLR and then dial the rest of the necessary information according to a specified procedure. For example, a WVLR, after a subscriber is connected to it, issues a tone and the subscriber enters the MIN and then the DN of a fixed phone. In the second access case, a subscriber dials a prefix (e.g., *44), the SSP issues a tone, and the subscriber dials the MIN. The SSP formats the MIN and the subscriber's fixed phone DN, which it knows from its association with a specific line, and sends the information using a specially defined protocol to a WVLR.

Figure 11.8 shows an example of how a call delivery to a subscriber registered as a wireline user would take place. The actions are as follows:

1. A caller originates a call by dialing the MIN—908-345-6789; the caller can be either a wireline or a wireless user.

2. An originating switch routes a call to an MSC that is associated with the MIN.

3. An MSC requests location information from the HLR.

4. An HLR sends a request for routing information to a WVLR, where the subscriber is currently registered.

5. A WVLR knows that this subscriber is currently at the fixed location served by DN 908-234-5678 and sends this number as the TLDN.

6. An HLR delivers a TLDN to a home MSC.

7. A home MSC routes a call to the SSP.

8. The SSP delivers a call to a fixed phone.

An interesting feature of the above process is that it is moving subscribers and service providers toward universal services. We will discuss this idea in more detail in Chap. 14. The requirements for the wireline registration service include:

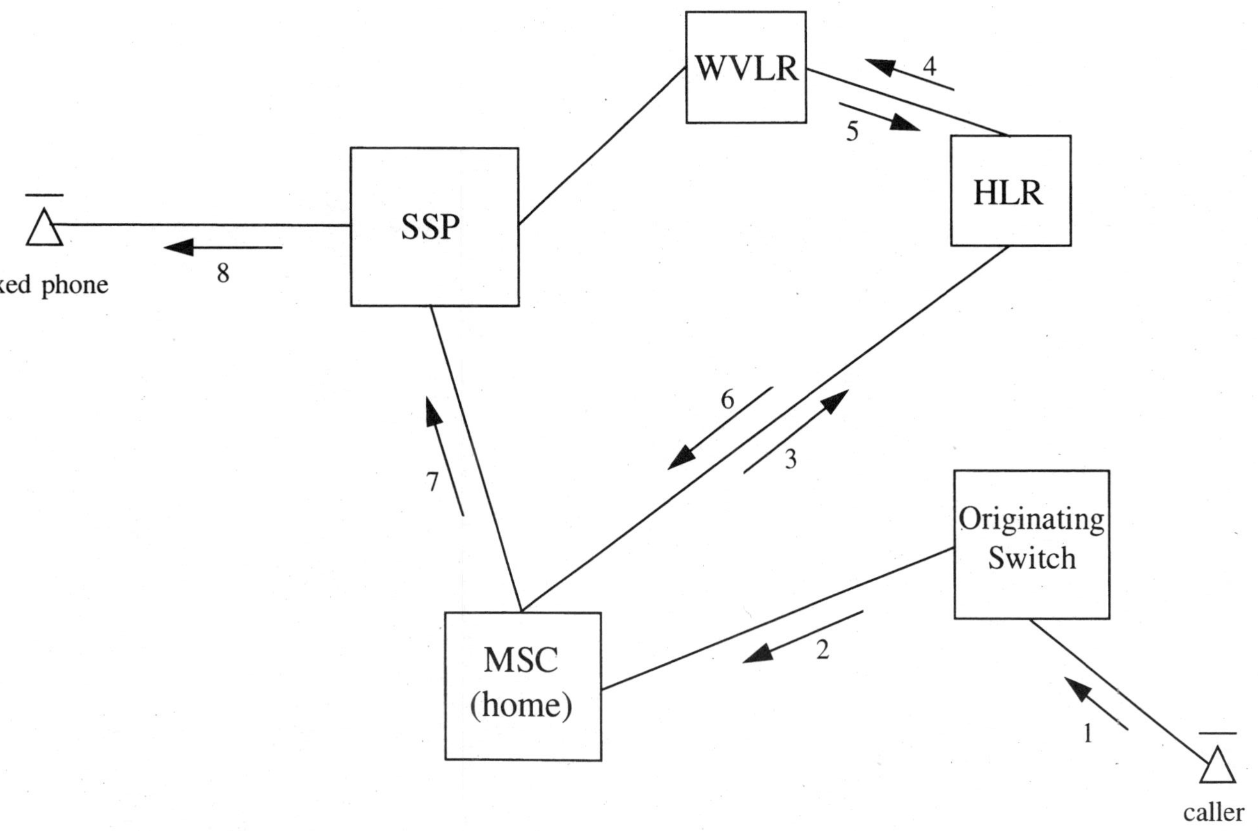

Figure 11.8 Call delivery to a wireless subscriber registered at a fixed phone.

- Development of a WVLR capable of receiving wireline registration; the modification of an existing VLR should be straightforward to accommodate this functionality.
- Development of an interface between an SSP and a WVLR.
- Development of a switch access feature.
- Possibly, development of a home base.

11.3 Fixed Wireless Loop[2]

The wireline network uses wires to connect its customers. This is a defining characteristic of a wireline network. We know this. So, what is the fixed (points to wireline) wireless (points to wireless) loop?

When we defined wireline networks in Chap. 3, it was mentioned that fixed wires present two problems: high cost of installation and maintenance and inability to support users on the move. It can even be said that fixed wireline access is a limitation for a greatly modernized network. Sometimes this limitation shows up as the inability of a wireline service provider to quickly respond to customers' requests, such as phone installation. Fixed wireless loop technology can help and, at the same time, provide a step toward a seamless wireline-wireless service.

A local loop, what we called a wire, actually consists of different segments (see Fig. 11.9). A feeder is a part of the local loop that connects a central office (CO) with a distribution center. The feeder plays the role of a multiplexer for traffic coming from different end points. A distribution is the network of wires that connect the distribution center with individual buildings or houses. Wiring within buildings is not part of the distribution.

Radio can replace either segment of the local loop. So, if a new wire cannot be installed or an existing one cannot be replaced for any reason, radio technology can be used in its place. This radio equipment would just replace a fixed wireline system. To the rest of the network—switches—no change would be visible.[3] That is why it is called *fixed wireless loop*. However, when it has been installed, it provides a potential for moving customers into the domain of mobility services.

Different technologies are available to serve in a fixed wireless loop arrangement. They range from Very Small Aperture Terminals (VSAT)

[2]Sometimes it is also called wireless local loop.

[3]An exception to this statement is the need to accommodate this new equipment within OSs for network management purposes.

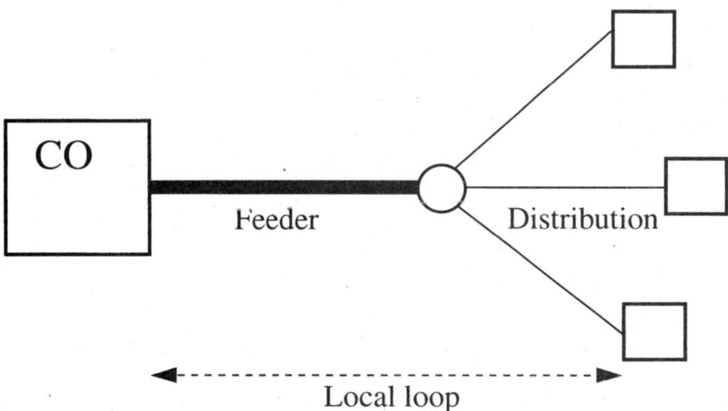

Figure 11.9 Local loop.

to various TDMA and CDMA systems. Application of each technology depends on the reasons for the use of radio in the loop and other conditions, such as the type of services that the loop would have to support. We will discuss these generic options for fixed wireless loops.

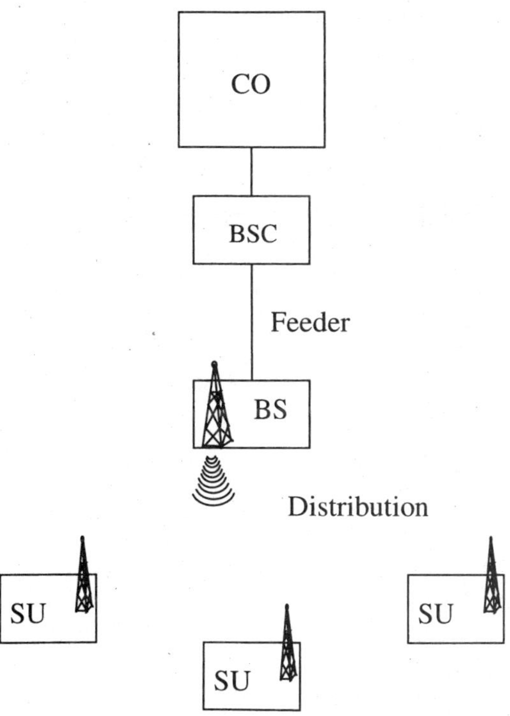

Figure 11.10 Fixed wireless loop architecture with wired feeder and wireless distribution.

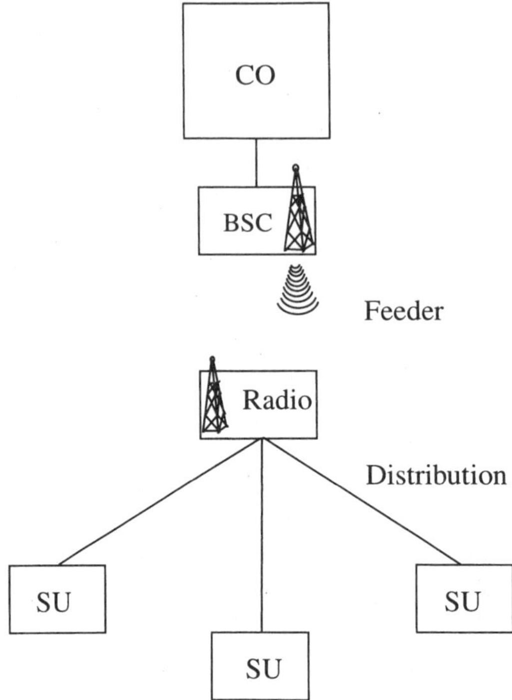

Figure 11.11 Fixed wireless loop architecture with wireless feeder and wired distribution.

The first option is the replacement of the distribution portion of the local loop with the radio. Figure 11.10 shows a radio BSC connected to a switch and to a BS (as it usually is in the wireless network). This is over the wired feeder. The point-to-multipoint radio is used between a BS and subscriber units (SUs).

The second option is shown in Fig. 11.11. In this option the distribution portion of the local loop stays unchanged, while the feeder portion is built using point-to-point radio.

In the third option, the point-to-multipoint solution replaces both the feeder and the distribution with radio (see Fig. 11.12). VSAT or other technology can be used for this option.

In all three examples the radio is not supposed to change anything for the subscribers it serves. The same services are usually supported, or at least that is the goal, as over wireline access lines, and voice quality is comparable. But what is interesting about this access architecture is that it is the first step toward mobility. The radio equipment installed in the local loop does not have to be equipped with any mobility-oriented functionality. The BSCs may be of a simpler type than their cellular and PCS counterparts to make them cheaper. However, these BSCs can be retrofitted with mobility functions, if they do not have them in the first place. This would change the fixed wireless loop from a substitute for a wire to a flexible access line. Given the right technology in the local loop, the subscribers served by this local loop can acquire handsets that can roam at slow speed within the coverage range of the BS, probably around their neighborhood. Actually, if the radio base stations are spread around an area, the subscribers can begin to experience the service that we previously defined as low tier.

Among the three examples given above, options 1 and 3 are good candidates for the type of evolution just described. In addition to the equipment shown in Figs. 11.11 and 11.12, a VLR may be needed to record subscriber profile information. If the area covered by this type of wireless access is significant enough, a centrally located HLR and a number of VLRs may be required.

Whether or not mobility services can be offered to subscribers connected to a wireless local loop depends on what kind of spectrum is

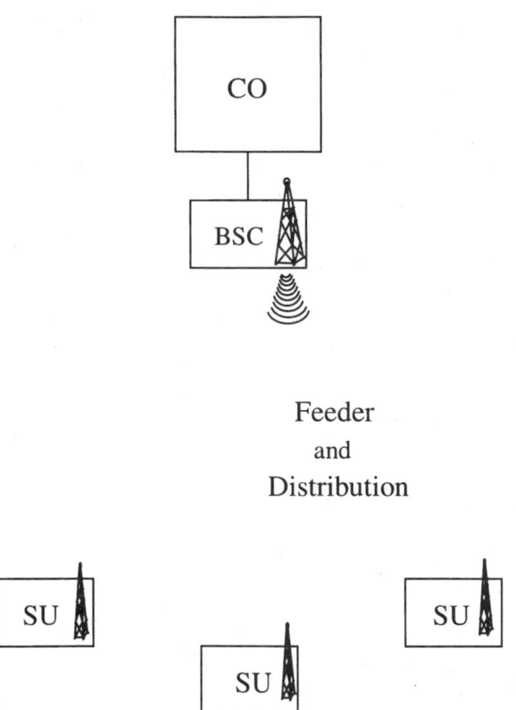

Figure 11.12 Fixed wireless loop architecture with wireless feeder and wireless distribution.

used for the initial purpose. Different spectrum can be used depending on technology. Examples are:

- Unlicensed PCS (at 1900 MHz)
- ISM bands (at 5700 MHz)
- MMDS band (at 2.6 GHz)

If an initial license has been granted only for fixed applications, a waiver is required to introduce new applications.

If the low-tier service provided by wireless local loop technology is merged with the high-tier service provided by high-tier technology, a multitier wireless service would be supported. Since the subscribers affected by such an arrangement still receive service at their homes (or wherever the original locations of the fixed wireless loop technology were), we can say that a degree of seamlessness has been achieved since the services provided to the subscribers will probably be seamless across the wireless and what is perceived to be the wireline component of the network.

The fixed wireless loop technology is becoming increasingly popular in countries where the wireline network infrastructure is underdeveloped. In Eastern Europe, Asia, and Africa it is faster and cheaper to introduce radio technology than to construct an outside wireline plant. By providing an economical alternative to a wired subscribers' loop, fixed wireless loop technology helps in the establishment of an important infrastructure that is a must in today's world.

11.4 Multitier Service

In Chap. 4 we introduced the multitier concept as a service differentiator. This concept can also be seen as a technique for providing seamless wireless services for customers using different radio technologies. Furthermore, it can incorporate elements of wireline networks, specifically a cordless phone with fixed access to the network. To recap, a multitier service is a service that supports subscribers using high- and low-mobility services depending on their location and speed of movement. A fast-moving subscriber would be served by a high-mobility system, and a subscriber who is stationary or moving slowly would be served by a low-mobility system since it is cheaper. We can extend this definition to include a third, even lower tier—cordless phones. The idea of a multitier service is still the same and it allows the subscriber a choice of access to make an overall service more economical. Figure 11.13 shows a conceptual model of a multitier service. A subscriber would be able to switch between tiers to gain

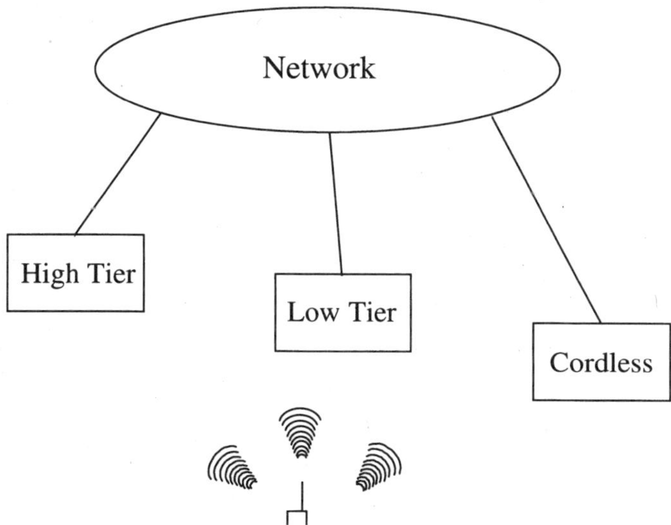

Figure 11.13 Multitier concept.

an economic benefit or a better quality of service (e.g., voice quality) or a service otherwise unavailable (e.g., a high-mobility tier for a phone in a fast moving car).

A multitier service requires the use of different technologies and as such, if implemented, supports the idea of a seamless network in its access part. Issues related to the lowest tier (cordless phone) have a lot in common with the wireline registration concept. Therefore, the discussion here will focus on the issues related to two tiers: low and high mobility. But first we need to look at the handset issue as it applies to multitier service.

In roaming between different radio systems the handset aspect of the service is important. A simple handset designed for a particular radio technology (e.g., AMPS) cannot be used with a different radio technology. Two obvious choices exist: Either a handset is designed to work with multiple radio technologies or multiple handsets are needed. Both choices have advantages and drawbacks. A multimode handset would hide technological differences of various radios from a subscriber, but it would be quite complex and would definitely cost more than a simpler handset. However, multiple single mode handsets would probably cost even more. The ultimate decision regarding the handset preference will depend on a particular customer's situation. If, for example, a customer clearly differentiates vehicular and pedestrian usage and can use a handset in either environment for a relatively long duration, two handsets may be better. If, on the other hand, a customer is in a constantly changing environment (in a car, walking, in a car again, then at the meeting, etc.), a single handset seems to be more appropriate. The variations in situations are many and not well understood yet. However, development of a multimode handset is going to take place since it is already happening for cellular services. Since the current cellular networks are based on AMPS and are migrating toward digital technologies, dual-mode handsets have been introduced. These handsets are not as complex as those required for a multitier service because the dissimilarities between AMPS and IS-54 TDMA are not as great as between, let's say, AMPS and PACS. In the former case both technologies are high tier, and in the latter case one is high and the other is low tier. Nevertheless, advances in wireless and digital technologies should make a multitier handset economically possible.

Now we need to discuss the issues of integrating low- and high-tier technologies into a multitier arrangement. As an example, we will use the architecture with AMPS as the high-tier technology and PACS as a low-tier technology.

Figure 11.14 shows a conceptual multitier arrangement that consists of wireless access via AMPS and PACS technologies. The two tiers have to communicate to provide a multitier service. The type of service to be provided will dictate the communications needs between the tiers. Conversely, the level of communications between the tiers will indicate which services can be provided. The simplest multitier service requires no communications between the tiers at all. A subscriber would have to sign either with two different service providers, one for each tier, or with a single service provider who offers both high- and low-tier services. In this case a subscriber would not receive any benefit from a multitier service beyond the ability to use a cheaper service when appropriate. The subscriber can have a single dual-mode phone or two phones. In either case getting on a different tier requires registration at that tier, and no handoffs between tiers are possible. This scenario does not seem to fit with the idea of a seamless network, but it is a possible scenario, similar to today's roaming arrangements when a subscriber roams into a "foreign" area and receives a service based on business agreements between two service providers.

A much more promising arrangement would allow some commonality of service across both tiers. A few things are important in this arrangement that would make it worthwhile. These are as follows:

- Ability to register only once
- Ability to choose a tier at will
- Ability to hand calls off from one tier to another
- Ability to receive the same services from both tiers

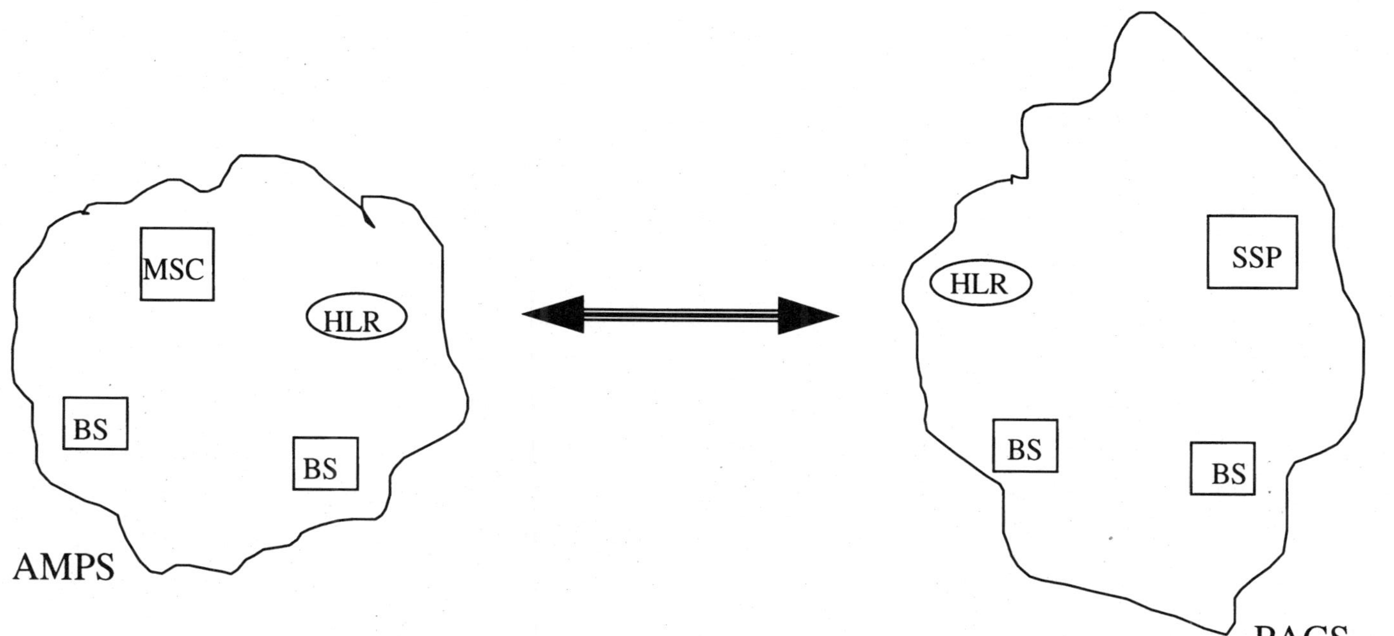

Figure 11.14 Multitier—AMPS and PACS.

11.4.1 Registration

It is important to realize that registration in a multitier environment is not the same as registration in a new location for a conventional roamer. In the latter case the registration signifies the location of the subscriber. In the former case it is not just the location which is being indicated but also which service level this subscriber can receive. Both high and low-tier systems may cover this subscriber in the same location, as illustrated in Fig. 11.15. The difference in coverage is based not on location but on other factors, such as speed. When a subscriber roams into a different registration area of the same tier, the reregistration is performed in the usual manner.

A multitier arrangement allows a subscriber to use a cheaper service. Hence, registration on two tiers simultaneously may be necessary whether two tiers serve the same area or not.

Two basic methods can be used for registration in a multitier environment:

- Single-tier registration
- Registrations at all tiers

A single-tier registration method requires a subscriber to register at one tier only. The selection of the tier is up to the subscriber, which gives the subscriber a choice and a guarantee that only one tier is ready to provide services. If the subscriber registers in a low tier, than only calls that the low tier can serve would be supported. If this subscriber begins driving, after the vehicle's speed reaches a certain limit, the low-tier system would not be capable of supporting calls to the subscriber's handset. In this situation a deregistration from the low

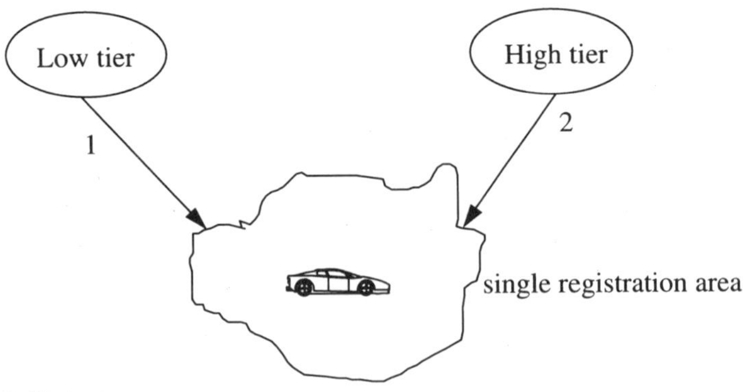

Figure 11.15 Single registration area for two tiers.

tier and registration in a high tier would have to be performed under the control of the subscriber.

This type of registration is simple to support since it does not require any changes in the procedures and operations of an HLR. If specific handsets are used for each tier, they do not require any changes either. If a single dual-mode handset is used, the handset has to allow the subscriber to indicate which type of service (high or low mobility) to request and has to attempt to register at the indicated tier.

A single-tier registration is simple and straightforward, but it does not allow an automatic tier selection by the network and, therefore, does not support the requirement of registering only once for the services provided by all tiers. Calls can be originated and terminated only using one particular tier at a time, and a subscriber has to watch for a changing environment to determine when to register in a different tier. As in most cases, the issue here is simplicity versus sophistication.

A much more sophisticated multitier service can be achieved with registration at multiple tiers. This would allow the network equipped with the information about a subscriber to automatically use the tier which is indicated by information stored in the subscriber's profile. This type of registration can be accomplished by two alternatives.

The first alternative requires a single registration request by a subscriber. This request can be issued on any applicable tier. When received by the chosen tier, a successful registration is placed in all tiers through a procedure which largely depends on the architecture of the network providing the multitier service. The important point is that by issuing a single request for registration, the subscriber is registered in all tiers.

The second alternative requires a subscriber to register at each tier individually. This is a more laborious procedure, but it would allow a subscriber to indicate a particular selection of tiers for a possible service. For example, in a four-tier arrangement a single registration request would register a subscriber in all four tiers, but a multiple registration request procedure allows the selection of, for example, only three particular tiers to be available for service. Therefore, if the subscriber does not wish to be reached on a high tier even when other tiers are not capable of delivering calls (e.g., when a vehicle is moving fast), no high-tier service will be offered and the more expensive air charges will be precluded.

11.4.2 Architectural alternatives

As mentioned before, what occurs when a single registration request is issued depends on the particular network architecture that supports the multitier arrangement. We will now spend some

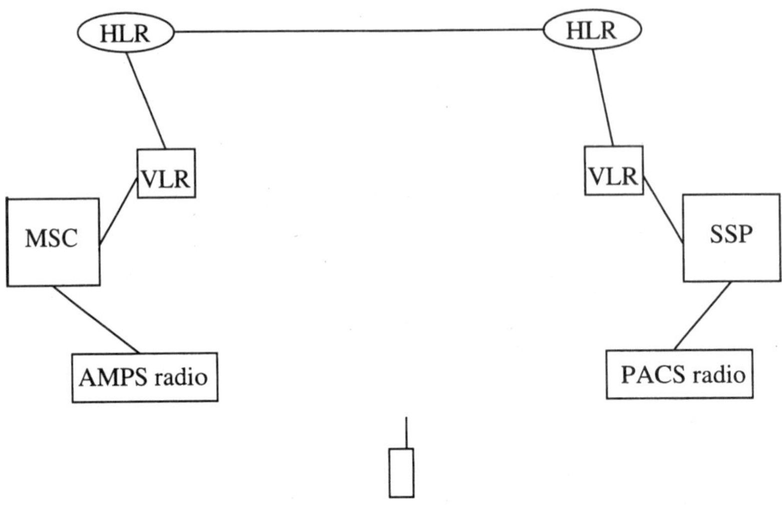

Figure 11.16 Multiple HLRs in a single registration request scenario.

time discussing the architectural alternatives by defining several scenarios.

The first scenario is shown in Fig. 11.16. Its distinguishing characteristic is an HLR in each tier. In this scenario a subscriber would register once via either AMPS or PACS, and the HLR would be aware of the current location of the subscriber. The HLR would inform the peer HLR (the other tier's HLR) of this registration because both HLRs need to know where the subscriber is. Such a capability does not exist today; so, if this service arrangement is to be deployed, a standard or a proprietary protocol for this type of HLR-to-HLR communication is necessary. The communication between HLRs may be quite complex since both must have exactly the same subscriber records and information about the subscriber's location, profile, and priorities. Synchronization requirements may be a prohibitive factor for this type of architecture.

While the previous scenario is based on the possibility of two service providers with developed networks and HLRs offering a multitier service, another scenario is a single network provider offering a multitier service. Here a single HLR is used. If a single service provider owns high- and low-tier radio access equipment, a single HLR makes a lot of sense. It is also conceivable that a single HLR could be used by two different owners of radio access equipment, but this possibility may be a bit remote. However, if this idea is combined with the generic access interface, it becomes a lot more plausible. Then the HLR may also be a part of the package offered, together with switching services, to two service providers by a public network. Regardless of ownership, the second scenario would be as shown in Fig. 11.17.

In this scenario the registration is accomplished as before with a single request. However, since a single HLR is used, there is no need for the inter-HLR communications. Instead, the HLR has to register the user on the tier where the user is not yet registered. The sequence of events could be as follows:

1. A subscriber sends a request for registration over the PACS air interface.
2. The HLR is consulted and the subscriber is registered in the PACS VLR.
3. The HLR registers the subscriber in the AMPS VLR.

At this point, the subscriber can originate and receive calls on either tier (using either AMPS or PACS). All services are delivered by the network based on the profile stored in the HLR, including priorities of radio systems to be used when a call is to be terminated at the subscriber's handset. The most economically minded profile would indicate that the first call delivery attempt is via PACS (low tier, cheaper service), and, if the subscriber is unreachable, the second attempt is via AMPS (high tier).

In this scenario each VLR "thinks" that the subscriber is registered only with it. The HLR has to track who is actually serving the subscriber. In the third scenario this burden is shifted to a new entity which we call a master VLR, or MVLR (see Fig. 11.18).

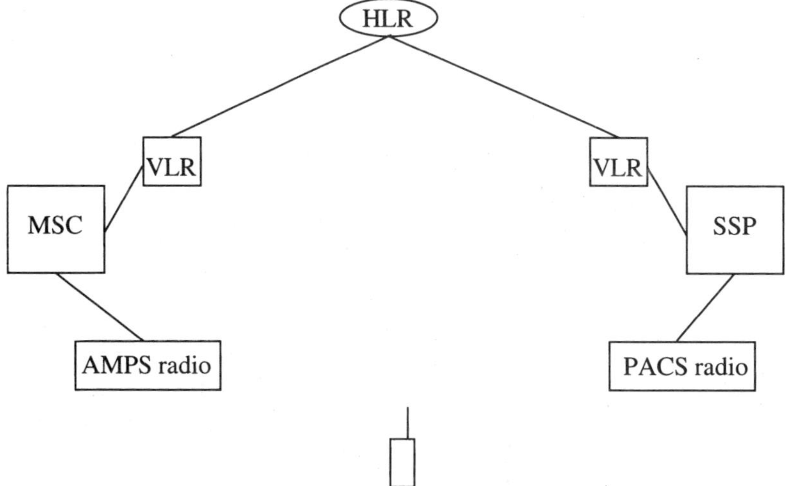

Figure 11.17 Single registration request method with a single HLR.

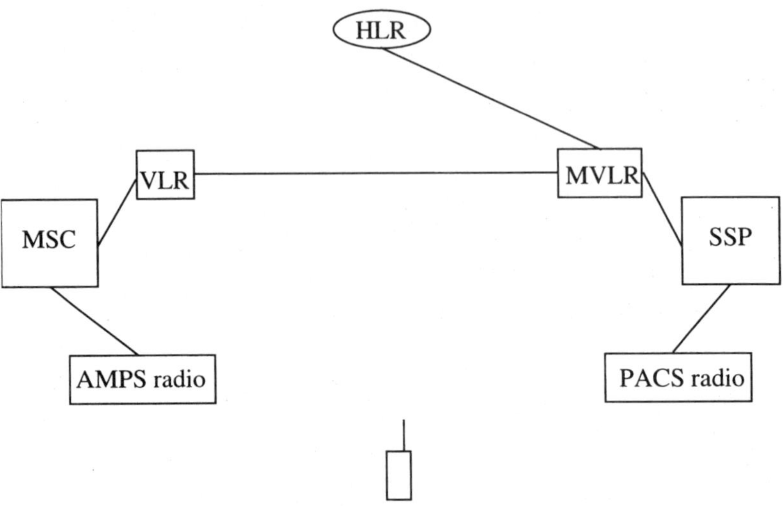

Figure 11.18 Multitier architecture with an MVLR.

With this architecture an HLR is only aware of one registration at a time. The fact that multiple VLRs at different tiers are capable of supporting a subscriber is hidden from the HLR by the MVLR, which now performs a function assigned to the HLR in the previous scenario. This function is to record the state of the subscriber vis à vis a particular tier.

The registration process can be represented as follows:

1. A subscriber sends a request for registration over the PACS air interface.
2. The HLR is consulted and the subscriber is registered in the PACS VLR.
3. The PACS VLR, performing the functions of an MVLR, attempts to register the subscriber with the AMPS VLR.
4. If the AMPS registration is successful, the PACS VLR reports to the HLR that it is performing as the MVLR.

All three scenarios that support a single request for registration in multiple tiers comply with the requirement to register only once. Procedures that a subscriber has to follow are thereby simplified. They also support the second requirement of choosing a tier at will through the profile process. The profile indicates the priorities of different tiers in regard to services and can include time- or location-based directives. Hence, the network can know that on Sundays no high-tier service is to be provided, while during the rest of the week, the high-tier service is to be offered first.

The second requirement is also clearly supported with an individual manual registration on each tier.

11.4.3 Handoffs

A multitier service can be defined in many different ways from the simplest subscription to two different service tiers to the most flexible service with sophisticated selection capabilities. For example, a selection of local tiers and their respective fees can be loaded into a handset. The handset will then automatically select the most appropriate profile based on the user's predefined needs. All registration and service options will be selected by a handset and used while in the area. This sophisticated multitier service that we are envisioning will also need handoff capabilities between different tiers. Handoffs between tiers would support calls in progress while transferring from one tier to another. Without these capabilities some calls would be dropped if the coverage could not be provided on the tier the call is on, or the economic benefit of using a cheaper tier could not be achieved.

The ability to perform intertier handoffs would be a good feature for multitier service. Two general types of handoff would have to be provided: upward and downward. The handoffs up the tiers are more important to maintain calls in progress. If a subscriber is served by the low-tier while walking toward the car, when the car speeds up, the call must be transferred to a high tier to avoid being dropped. On the other hand, if a subscriber is on a phone while driving at high speed, becoming stationary does not imply that the call cannot be sustained. It just means that a cheaper form of service cannot be obtained if high-to-low-tier handoffs are not supported.

Intertier handoffs would not be easy to implement. At this time, there are no special efforts to make this happen. This kind of a process would require significant resources to achieve workable implementations. Obviously, a multitier service can be provided without intertier handoffs, but they would make the service more attractive (if the price is right, of course).

The issues of intertier handoffs depend on particular technologies supporting various tiers. Among different technologies capable of supporting high and low tiers are CDMA and various forms of TDMA (e.g., GSM, PACS, IS-54/136). If a multitier network is built using different technologies, a handoff between them would be a lot more difficult than if similar technologies were used. Two elements are important in making handoffs work: radio and network procedures.

Radio procedures are used to measure the quality of voice, acquire a new channel, request a handoff, and drop an old radio channel. A handoff between a CDMA and a TDMA system would be difficult. Network procedures are used to transfer a user from one network element to another (e.g., an MSC or a BSC). These procedures are also different, for example, between PACS and GSM, although both are TDMA systems. Again, the intertier handoff would require changes in the established procedures for each technology. The solution for these types of handoffs may be in a new network-based function that would keep track of differences between radio systems and map different procedures so a handoff can be accomplished. This process would add time to the handoff procedure. The decision regarding the plausibility of an intertier handoff feature as part of a multitier service would have to be made based on market need. Unfortunately, it is very difficult to estimate a market need when even a rudimentary multitier service is not available, and people do not have experience with this kind of technology.

11.4.4 Services

The ability to receive services in a transparent manner across different tiers is another requirement that we set for multitier service. Its fulfillment requires different tiers to be capable of supporting the same services independent of each other. If one tier cannot support a service that another tier can, obviously this service cannot be transparent. But let's assume that the same services are supported by all tiers. In this case, although each individual service would require a detailed analysis, in general the uniformity of services should be preserved. They are controlled using the same HLR (with an exception of a scenario with multiple HLRs), and, if individual tiers are capable of supporting a particular service, they could be supported the same way in different tiers.

To observe the operations in a multitier arrangement we can examine a call delivery process (see Fig. 11.19). In this scenario a user at a fixed phone (shown connected to the originating switch) is calling a wireless user. The next steps in the call delivery procedure could be as follows (some operations are simplified for compactness, e.g., a switch and a VLR are shown as one entity):

1. An originating switch (this represents the first switch encountering the call which is capable of accessing the HLR) requests location information from the HLR.

2. The HLR checks the destination number and finds that this user is served by multiple tiers; the HLR checks the user's profile and identifies that the low-tier service has to be used first, if it is not possible, the high tier is used. The HLR sends a request to the VLR where the user was last registered (which is with PACS).

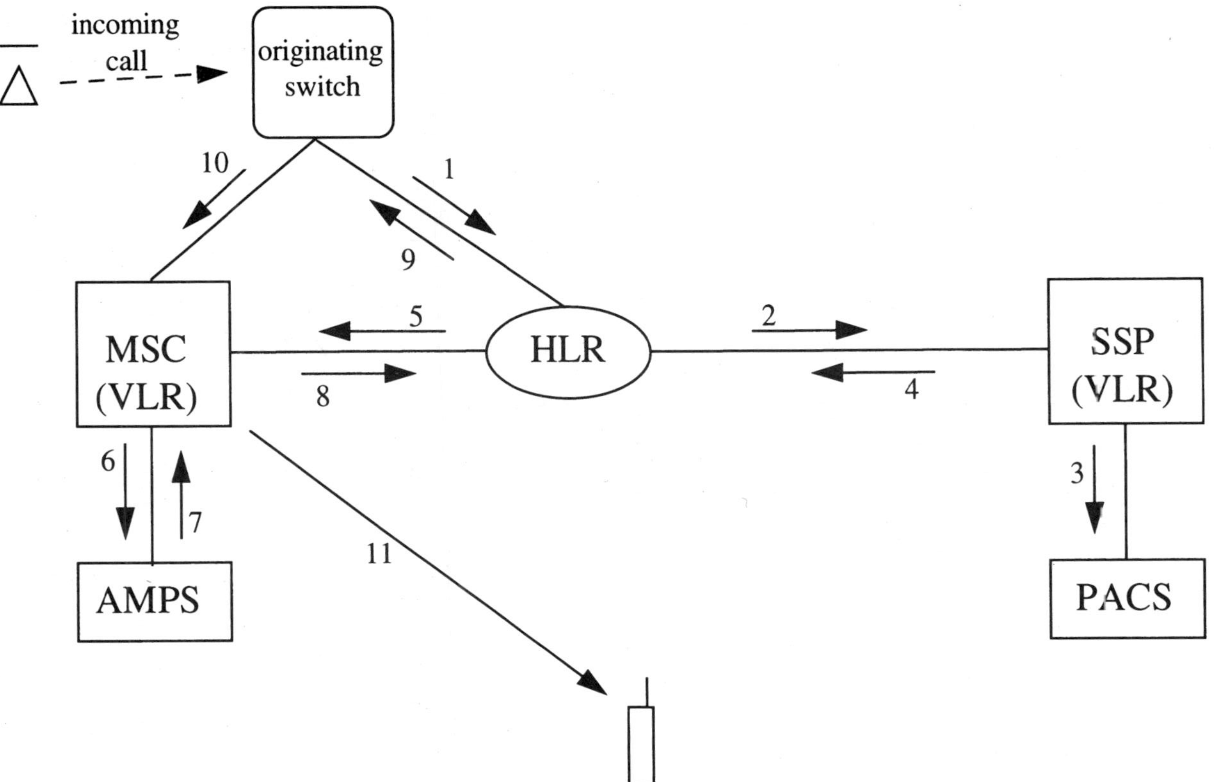

Figure 11.19 Call delivery in a multitier network.

3. The PACS VLR requests an SSP to find the called user. The PACS radio is ordered to perform paging.
4. The destination does not respond (e.g., the user is in a fast moving vehicle and the system knows this by measuring the number of radio port changes in an interval); the PACS VLR rejects the HLR's request.
5. The HLR sends a request to the AMPS VLR.
6. The AMPS VLR requests the switch to find the called user. The AMPS radio is ordered to perform paging.
7. The destination is found; the result is reported to the MSC.
8. Information on the location of the handset is sent to the HLR.
9. The HLR responds to the originating switch with a TLDN.
10. A switch routes a call to the AMPS MSC.
11. A call is delivered to the destination handset.

Views about a multitier service, like many other issues related to seamless networks, are many and varied. They differ among different industry members, which is not surprising. Whether such a service will be accepted in the marketplace depends on its correct definition. In the end, the winning service will do what customers need. Since the entire PCS market is just developing, it is not easy to understand these needs. Therefore, some people construct a multitier service using tiers outlined here, and others include additional tiers in the arrangement, such as unlicensed and satellite-based ones. Regardless of specific tiers defined or technologies used for these tiers, the issues that require solutions are similar to those discussed here: registration, network architecture, handoffs, services, and others. A possible multitier scenario of the future could include a high tier, a low tier, a cordless phone, a wireless PBX, and a satellite access (used when a user is really away from crowded places).

11.5 Merging Network Access

The various network access techniques discussed in this chapter can be combined to provide a higher degree of seamlessness. For example, the combination of a generic network access interface with a multitier arrangement and with wireline registration would provide quite a seamless environment for a service provider and for subscribers. The generic network access interface would aid a service provider by simplifying operations of different radios; the multitier service would sup-

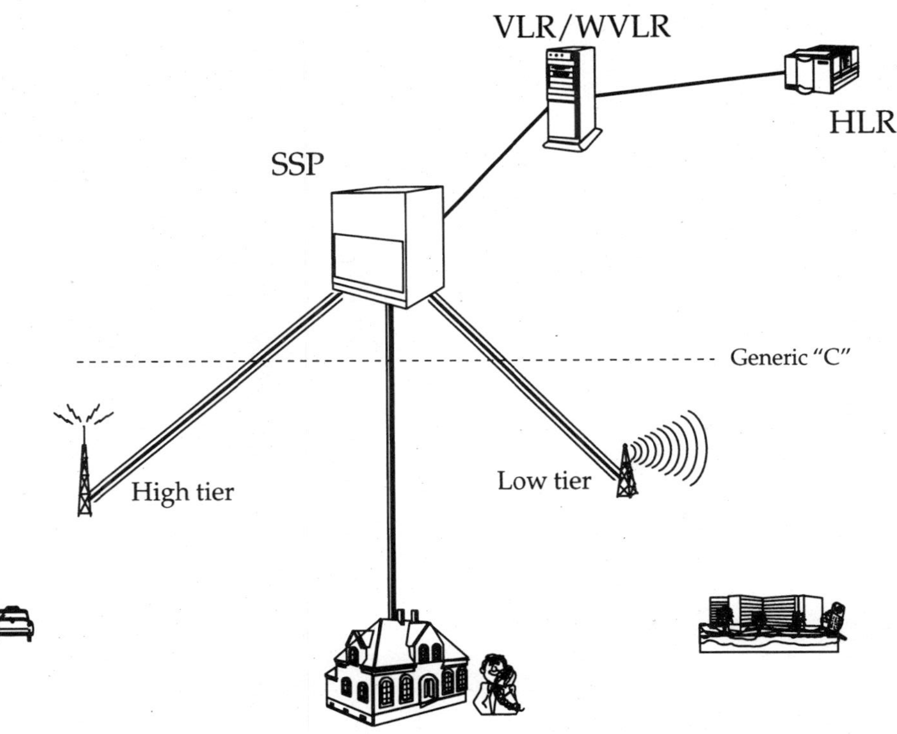

Figure 11.20 Combination of access techniques.

port subscribers' seamless services over different access technologies and provide an additional benefit of an economical choice; the wireline registration would extend a seamless service into a wireline environment and would also allow a subscriber to use a cheaper wireless service. The entire arrangement is shown in Fig. 11.20.

Chapter

12

Network Management

A telecommunications network of any type can be well designed, can be implemented with the best equipment, and can offer excellent services, but none of this will matter if network management is failing. This is an obvious fact because only through the use of proper administration, network monitoring for different purposes, and correct billing can a network operator be profitable and plan for the future.

A telecommunications network, either wireline or wireless, usually contains a number of different operations systems (OSs) performing different aspects of network management. These OSs are built according to the particular requirements of a given network operator and the requirements are usually not the same. Interoperability issues can surface in network management due to different reasons, such as the following:

- A need to provide a common service offered by multiple providers
- A need to build a network using different types of equipment
- A need to build a network by combining dissimilar subnetworks

Different OSs are usually deployed for different types of functions, and the resulting arrangement of OSs and interlinking facilities is quite complex. To deal with this situation a particular concept has been under development in standards bodies. The goal of this development is to create a systemic environment for network management. This concept is called the Telecommunications Management Network (TMN). Instead of trying to identify potential problems with some particular systems, it is better to discuss the TMN: an approach to network management that promises a simplification of the complexity and a multivendor environment through development of standards.

12.1 TMN

TMN is an approach to the network management issue that addresses interoperability among various operations systems, workstations, and managed network elements (MEs). It is being developed by ITU-T and is supported by many countries, including the United States. According to ITU-T draft recommendation M.3010,[1]

> The basic concept behind a TMN is to provide an organized architecture to achieve the interconnection between various types of Operations Systems (OSs) and/or telecommunications equipment for the exchange of management information using an agreed architecture with standardized interfaces including protocols and messages.

TMN is a network of various operations systems that communicates with a telecommunications network using standard interfaces. Figure 12.1 shows this relationship. This is a physical representation of the role TMN is playing, which is not different from a typical nonstandard network management. What is different here is that all data communications are handled by the same network, which already provides a certain degree of commonality. But this is not all. The idea of TMN is derived from principles of functional independence that we already discussed when we talked about Open System Interconnection (OSI).

The concept of TMN is also built around the idea of functional separation and building blocks, and it is applied in two areas: definitions of network management functions and the definition of the application protocol that performs communications tasks between various OSs and network elements.

12.1.1 Functional architecture

The activities of a TMN are partitioned into five layers. Each layer focuses on a set of activities assigned to it. Layers "communicate" with each other to complement each other's activities. The layers are defined as follows:

- Business management layer (BML) functions are responsible for overall TMN support. BML is associated with the goal-setting business functions, such as:

  Supporting the decision-making process for investment and the use of telecommunications resources

  Supporting budget planning

  Maintaining data about the entire enterprise

[1] The M.3010 status document, version 6. June 1995.

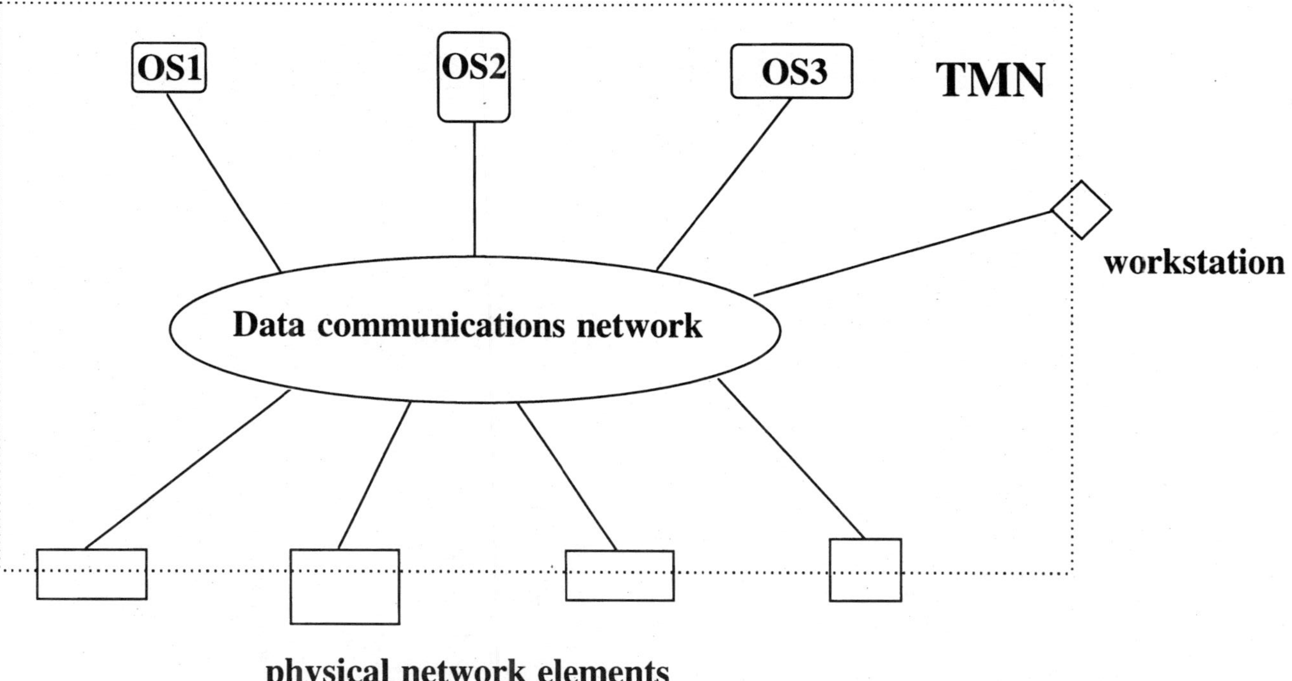

Figure 12.1 TMN.

- Service management layer (SML) functions are responsible for contractual aspects of services. Examples of functions it performs are:

  Contacts with customers

  Interactions with other service providers

  Reports of usage for billing purposes

  Maintenance of service data

- Network management layer (NML) functions are responsible for the management of the network. NML performs network coordinating functions such as:

  Maintaining the total network view, which includes all network nodes and links

  Provisioning and modifying of network capabilities for the support of services or customers

  Maintaining network capabilities

  Maintaining network data

- Element management layer (EML) functions are responsible for the management of individual network elements or groups of network elements; EML's main functions are:

  Control and coordination of individual network elements

  Control and coordination of subsets of network elements

  Maintenance of data regarding managed network elements

- Network element layer (NEL) is a representation of managed network elements. NEs are managed by the other four management layers and perform functions which are part of their behavior (e.g., switching, data collection).

Network management functions are varied and in TMN they are grouped into five areas:

- *Performance management.* Deals with performance monitoring, analysis, and management control. In general, it combines functions related to the behavior of the telecommunications network and its elements.

- *Fault management.* Deals with fault localization, alarm surveillance, reliability and survivability, quality assurance, trouble administration, and fault correction. This is basically an area related to maintenance.

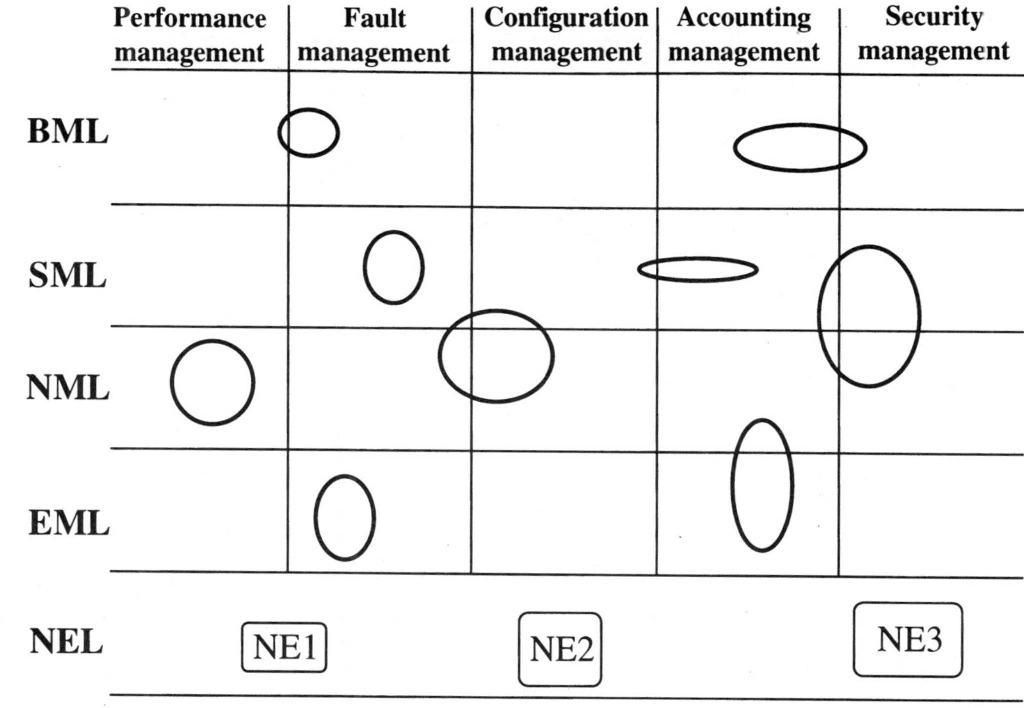

Figure 12.2 Functional TMN architecture with possible physical OSs represented by circles and ovals.

- *Configuration management.* Deals with network planning, network engineering, and provisioning. In general, these functions are related to the network setup.
- *Accounting management.* Deals with tariffs, billing, usage measurement, and pricing. In general, it can be characterized as accounting-oriented.
- *Security management.* Deals with issues such as fraud prevention and detection, security administration, and recovery. Obviously, all aspects of security are in this domain.

The functions in each of these areas are applied at every layer. Figure 12.2 shows how different functions relate to activity layers. Any intersection of a layer and a functional area represents a logical operations system. A collection of such systems can be developed as a physical operations system. It does not have to be built within the precise functional boundaries shown in the matrix; it can contain functionality belonging to different intersections. However, such an OS would use standard interfaces and standard descriptions of managed objects defined according to TMN. The definition of managed objects and the use of the same object descriptions by all involved systems is one of the key advantages of TMN. Examples of managed objects are calls, trunks, line cards, specific customers records, and so on. Not all necessary objects are currently defined and, therefore, their behavior is not described in standard terms. Only when a necessary set of objects and their behavior are standardized one can speak of migration to TMN for a particular set of management functions.

12.1.2 Protocol architecture

The protocol between OSs and NEs is also standardized by ITU-T. It is an application service element (ASE) and its name is *Common Management Information Service Element* (CMISE). CMISE consists of two parts: Common Management Information Service and Common Management Information Protocol (CMIS and CMIP). Figure 12.3 shows the protocol architecture for CMISE.

CMIS specifies services that CMISE provides to applications. These services are accessed via primitives which an application uses to invoke CMIS operations. Seven generic services (operations) are currently defined. These are:

- *M-Event_Report.* Reports an event related to an object
- *M-Get.* Requests retrieval of management information related to an object

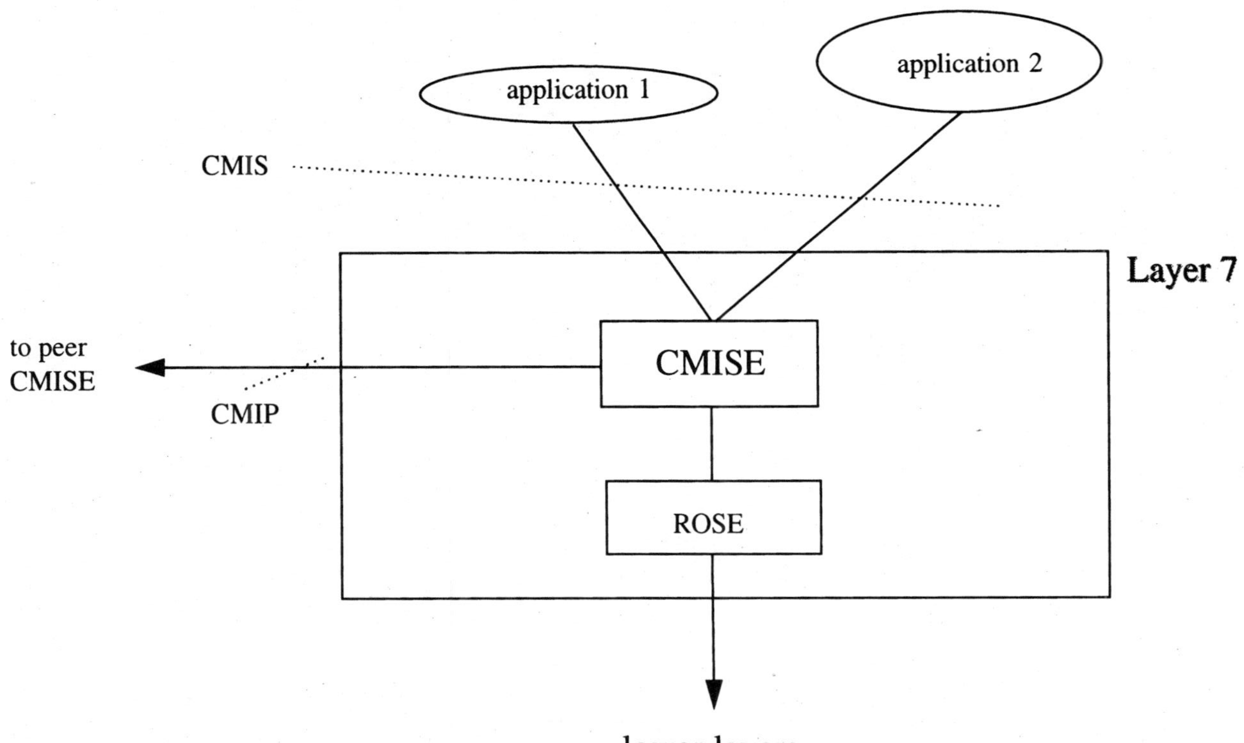

Figure 12.3 CMISE.

- *M-Set.* Requests modification of information related to an object
- *M-Action.* Requests an action to be performed
- *M-Create.* Requests the creation of an instance of an object
- *M-Delete.* Requests the deletion of an instance of an object
- *M-Cancel-Get.* Requests cancellation of an outstanding invocation of the M-Get service

These operations can be applied to any appropriate managed object.

CMIP is a protocol which is used to communicate with the peer CMISE application through the rest of a protocol stack. It creates specific messages that are communicated with the peer application. CMIP uses the Remote Operations Service Element (ROSE) as a mechanism for delivering messages to remote entities and receiving messages from those remote locations.

The application of TMN to the issue of interoperability is quite simple. If TMN can be used in the OSs that will operate in the network, the interoperability issue will not exist from the network management point of view. The question is obviously how to achieve it. And the solution depends on whether new equipment is being procured or existing equipment is going to be used. In the case of new equipment the situation is simpler since conformance with TMN standards can be requested from equipment vendors. When existing equipment is to be used, the process of modification to make existing OSs conformant to TMN may not be simple.

TMN is still being developed by standards bodies such as ITU-T. When fully developed it will assure seamlessness in the area of network management through the use of common interfaces and common managed objects. Some standards are already available, for example, for fault management. And, of course, the use of CMISE provides a common interface that can accommodate future TMN standards. The work continues, but planning for TMN could be performed without complete results in all functional areas. An appropriate TMN foundation would allow easier transition to more of the standardized functions when they become available.

12.2 Billing

Network activities directed toward measuring usage and generating bills is covered by TMN. We focus on this issue and not on the other network management issues because billing activities have unique implications for a business and place demanding requirements on NEs. Some issues are related to local regulations, and others affect the customer's perception of services. Therefore, some discussion of related issues is warranted.

To start, *billing* is not a precise term describing network activities leading toward the creation of bills to subscribers. *Accounting management and billing* is more appropriate—a definition which points to two aspects of what this chapter calls *billing*. The first aspect is accounting management: the collection of usage measurements and their transport to the appropriate billing system. The second aspect is an analysis of the accounting data and production of bills. Integration of these activities among different types of networks requires careful analysis of current differing approaches.

First we need to address usage data generation and collection issues. In general, usage is measured in the most appropriate places and then passed to the appropriate network entities for collection. We need to distinguish two aspects of these activities: record generation methods and formats of the usage data records.

A record generation method can be either centralized or distributed. In a centralized method all record collection takes place in one location—usually in a switch. For a given call, all necessary data are sent to the switch responsible for accounting management on this call. This centralized NE then transports the records to the appropriate operations system. Wireline networks and Global System Mobile (GSM) networks use this accounting method.

A distributed record generation method implies various network entities sending information to the appropriate operations system. Existing cellular networks use this method because of the flexibility it allows in using various network components to collect information about a call, especially when a call is handled by more than one network.

The second aspect of data generation and collection is the format of records used for this purpose; currently three formats exist: Automatic Message Accounting (AMA), Data Message Handling (DMH), and GSM. The AMA format was created for the wireline networks. The DMH format is a new approach developed for cellular networks. The GSM format has been defined specifically for GSM networks. Although the basic information elements of these formats are similar, some discrepancies exist. Some data exist in one format but not in others. For example, *billing identification number* exists in DMH but not other formats; *billing digits* and *alternate billing digits* are mandatory in the AMA format, optional in DMH, and do not exist in GSM. In addition, similar data elements in AMA, DMH, or GSM may have different formats.

So far we mentioned different methods used in cellular, wireline, and GSM networks. What about Personal Communications Services (PCS) networks? A subcommittee of T1, T1M1, is actively working to define requirements for accounting management in PCS. These requirements are based on the available information about existing systems and PCS plans. Essentially, another—fourth—approach is being created. Obviously such an important aspect of network management has an impact on the level of integration among different networks. The differences in formats and collection methods create a different billing environment for each of these networks. TMN can help by defining generic records and interfaces over which record collection would take place.

Production of bills to subscribers depends on the data accumulated for this purpose. Divergence in the type of data collected for billing is later translated into the different bills, which are sent to the wireline and wireless customers. While today even different wireline network operators use different formats for customer bills, it would be beneficial from the customer's convenience point of view to harmonize all bills. This is the second aspect of billing-related activities, and the result of this activity is visible to the customers.

Roaming presents another set of problems in wireless networks because of the need to correlate activities in "foreign" networks with information in the home network for each roaming user. Regulations in the country which is a "foreign" territory for a subscriber will play a role in how a service provider performs measurements activity.

Agreements among various network operators are necessary to set up a billing collection scheme. It is important to agree on who is being charged for which portion of a call (e.g., originating air, terminating air, call forwarding to a distant phone number). The mechanism for billing collection will depend on these agreements since they will specify where usage data are to be sent. As an example, in the wireline networks a calling party usually pays for the call, but in the wireless networks the air portion of the call, which is expensive, is usually charged to the owner of the handset used on that air interface. However, it is possible for a calling party to be responsible for such a charge. This depends on the service and may depend on the regulations in a particular jurisdiction.

Chapter

13

Seamless Services

One of the main goals of building seamless networks is to provide seamless services—the types of services that do not change their behavior when a subscriber crosses network boundaries. This applies to the boundaries between wireline and wireless networks, as well as to the boundaries between various wireless networks. This is sometimes referred to as feature transparency. While other expected positive results of seamless networks (operational savings, competitive advantage, simpler evolutionary path, etc.) are also important, seamless services directly affect customers. The convenience of uniform services will increase their usage and generate more revenues.

Making services seamless requires a detailed analysis of each service and its implementation in the wireline and wireless networks. Many times a service can be seamless if the interconnection of networks includes the same signaling system Signaling Sysem #7 (SS7). Other services require specific adaptations to the needs of the interoperating networks. We will look at some examples of services in this chapter but first will discuss roaming.

13.1 Roaming

Roaming is a service offered by wireless networks. It allows subscribers to receive services when away from their home area. When a subscriber roams into a "foreign" territory, the local network recognizes that this subscriber can be served because of the roaming arrangement between the two service providers and the subscription of this particular user to the roaming service. When a user roams, the selection of services that this user can receive depends not only on the user's subscription but on the capabilities of the network which currently serves this user. Even if solutions for all interoperability issues are found, the other necessary condition is to upgrade all appropriate network elements with new capabilities that support the same services. If a specific Mobile Switching Center (MSC) or home location register (HLR) does not have a particular capability required to support a service, regardless of the subscription and the available technical solutions, the user covered by these systems cannot expect a seamless service.

In addition to being a service, roaming also generates demand for feature transparency of all other services. This means that roaming has to be supported as well as the interworking of networks so that all supplementary services (e.g., call waiting, voice messaging) are supported. Roaming creates demand for this kind of seamlessness because it allows people to move across service providers borders with the expectation that the same services are available.

13.2 Availability of SS7

Another general issue related to seamless services is availability of SS7 in different networks. Many services require passing of certain parameters from the place of call origination to the place of call termination. For instance, calling party number is a required parameter in many custom local area switching services (CLASS). If end equipment (e.g., switches) is capable of using such parameters in the way appropriate for a particular service, but SS7 is not used to transport these parameters, the services that rely on them cannot be supported. This aspect of service seamlessness does not need an analysis and a technical solution. It needs a modern signaling system to become part of a particular network.

One such service is Calling Number Delivery (CND). It consists of the delivery to the subscriber of a calling number. The calling party number can easily be transported from the originating switch to the terminating switch in the SS7 Initial Address Message (IAM). Figure 13.1 shows a CND service model. The call is originated by the wireless user identified by DN = X. This identifier is passed in the SS7 IAM as the calling party number. It is delivered to the wireline user according to the service subscription.

Another example is a service called Calling Name Delivery (CNAM). It is similar to CND but delivers the name of a caller instead of the calling number. It works by mapping calling numbers to names associated with these numbers (e.g., if Joe Bloe's telephone number is Z, a special database will contain an entry Z = Joe Bloe). It also requires SS7 to be deployed so a calling party number can be delivered to the termination switch. But, in addition to this requirement, it also needs CNAM databases that retain all those mapping tables. Access

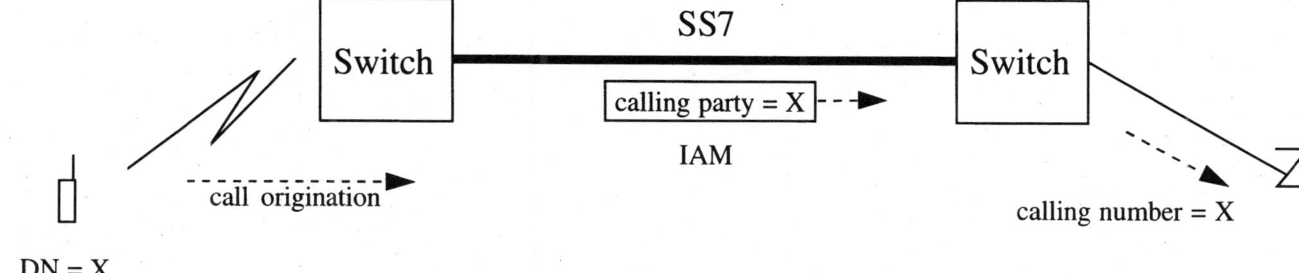

Figure 13.1 CND model.

to these databases is required from the terminating network, whether it is wireless or wireline. A seamless CNAM service would require equal opportunity for database access to all interested parties (wireline or wireless). Such access, beside the business agreements, requires (1) commonality of applications controlling the services and (2) the protocols that support the necessary communications.

Other similar examples of services are Automatic Recall, Call Screening, Selective Ringing, and so on. Their common feature is the need for the calling party number to be delivered to the terminating switch. If the delivery is possible, that is, if an interface supports this need through availability of SS7, the rest depends on the systems' capabilities. That is, the systems-related issues must be solved to offer a service (see Chap. 5 for the description of this analysis method).

An analysis of each service would generate a list of requirements that need to be followed in order to make a service seamless. Although these requirements are different for various services, they are all based on the issues we already discussed, such as network architecture, application protocols, interconnection, etc. In many instances all that would be required is to deploy available technology (e.g., SS7); in other instances new solutions or standards may be needed.

Another side to seamless networks is the need to provide the same user interface. If different sequences of events, such as dialing a sequence of numbers, are required to invoke the same service within different domains (wireline and wireless or different networks of the same type), the users' willingness to use such services may be diminished. Various inconsistencies in network procedures, such as issuing different tones at the same stages of a service, would have the same effect. All these differences add to diminished service seamlessness and become a barrier to a greater usage.

Providing seamless services is key to justifying seamless networks. Potential service providers need to understand the services they *must* make seamless and the services they *should* make seamless. The analysis of each service from the system and interface point of view would show what needs to be done to make a seamless service.

Chapter

14

Universal Personal Telecommunications

The concept of Universal Personal Telecommunications (UPT) appeared in the international arena a few years ago. It gave a boost to the ongoing development of international intelligent network (IN) standards by clearly identifying IN as the platform on which UPT as an application would be built. UPT would use the IN-defined functions to create services and would use IN-defined standards to promote the widespread penetration of UPT services. In fact, our investigation of issues that affect seamless networking is the investigation of how to approach the UPT age. The American personal communications services (PCS) concept is also partially based on UPT because many people view PCS not only as a particular spectrum offering but as a promise of universal (within the U.S. borders) personal communications. In short, it can be stated that achieving seamless networking is the same as rolling out UPT in all parts of the world. While the work on UPT-related technologies and standards is ongoing, the concept can be described, and some standards, such as IN, are already available.

14.1 UPT Concept

The concept of universal communications with no boundaries and the ability to reach people anywhere in the world is not difficult to comprehend. It is just difficult to implement. Difficulties usually do not stop people from pursuing hard-to-achieve goals, and they should not stop us on the road to UPT either. Actually, the state of some technologies is such that it would not be difficult to see UPT services, some of them in an immature fashion, appearing here and there, stimulating interest, gaining a foothold, and then growing. Among the technologies that are necessary to support UPT are microelectronics, computers, computing techniques (e.g., database management, information retrieval, and parallel processing), transmission technologies, and common worldwide standards.

The development of IN standards is key to achieving universal personal services for two reasons. First, UPT services will require a computer-based platform for the variety of applications that it promises. All kinds of information would have to be stored in these computers to make complex services possible. And the same computers would have to process a lot of information by mapping various data elements, finding appropriate entries, and analyzing information. These computers would also have to communicate with each other and with other network elements, such as switches. The IN platform has been specifically designed for these kinds of applications. It identifies how different service building blocks are to be partitioned to be used for different services and how various network elements (databases, switches) are to communicate.

The second reason is that the international standard is necessary for the worldwide interoperability, and the IN platform defined by ITU-T is just that—a standard. Without a standard way of accomplishing a task, there is no hope for UPT to become globally supported. Even if technology used in different parts of the world is similar, without standards there will be enough differences to make universal seamless services impossible. Only a standard, formal or defacto, can help with the goal of universal services.

In high-level terms, UPT is an ability to reach a subscriber anywhere in the world, distribute information to this subscriber in a predefined manner, and allow the subscriber to change a subscription profile at will. This short definition translates into three categories of services: personal numbering, flexible call routing, and dynamic subscriber modifications.

Personal numbering

The assignment of a personal number to a subscriber and the ability to locate the subscriber are key to the UPT service. A subscriber is not to be limited by the constraints imposed by the existing technique of identifying not a subscriber but a particular terminal that the subscriber is using. With UPT, any device can become a personal communicator for a person who identifies him- or herself to the network as using this particular device. This is conceptually done by registering with the network on a device through some sequence of dialed characters that would include a personal number. When this registration occurs, all calls destined to this number would be routed to this particular device.

Flexible call routing

When a subscriber has a number of different devices, such as fax machines, computers, and voice mailboxes (these can also be personal computers), the subscriber can direct information delivery to a particular device by specifying the conditions of delivery in the profile. The profile may say that all calls are to be routed to a fixed phone at this particular location, unless... Here a long list of defaults can be specified. For example,

- Route calls from Japan to a fax machine in Texas
- Route calls from number ZXC to a voice mailbox and inform me immediately
- Route calls after 3 p.m. to my mobile telephone
- Route calls after 5 p.m. every Sunday to my personal computer as e-mail
- Route calls in June to my office in London

The important characteristic of this flexible routing is that all devices have the same telephone number.

Dynamic subscriber modifications

Dynamic modification by a subscriber of a profile can be made on a temporary or permanent basis. Temporary means call screening abilities that allow a subscriber to decide what to do with a call when some information about this call is available (e.g., a calling name). Permanent means the ability of a subscriber to alter a profile at will. This flexibility also pertains to billing options (e.g., where to send a bill, and what currency to use).

The UPT concept can be summarized as one number per subscriber, reachable anywhere in a form specified by the subscriber using flexible profile options.

14.2 UPT Standards

Currently, beside the reliance on the development of IN standards, work is also underway to specify UPT services in ITU-T. The list of services for UPT is a long one. It covers the existing Integrated Services Digital Network (ISDN) services and also includes new services. Many of these are familiar services, but the important point to remember is that even these familiar services will take an additional flavor because of the UPT service mobility approach. Service mobility can be described as an ability to invoke the subscribed services in any part of the world regardless of the particular network serving the subscriber at any given moment.

The work on UPT services is still going on in ITU-T Study Group I. The current documentation includes the following services (this is a partial list to present the type of services addressed by UPT):

Existing services

- Call waiting
- Call hold
- Conference calling
- Calling line identification presentation
- Calling line identification restriction
- Calling name identification presentation
- Calling name identification restriction
- Call transfer

New services

- Connected user identity presentation; an ability to define a name or a number presented to the calling party instead of the one supplied by the supporting network
- Connected user identity restriction; the ability to restrict the presentation of the subscriber's identity to the calling user
- Call importance indication; the ability to indicate to the called party the level of call importance
- Selective call forwarding on different conditions, such as *Busy Subscriber* or *No Reply*
- Call screening services

14.3 Toward Worldwide Networking

This may sound utopian at this time, but the technology of tomorrow *should* sound strange today. It is hard for most people to imagine a world that allows people to maintain their personal numbers and services anywhere, and to have services of a science fiction nature, such as telling a device what to do instead of typing something into it. But some of these services are slowly being introduced right now, and a lot

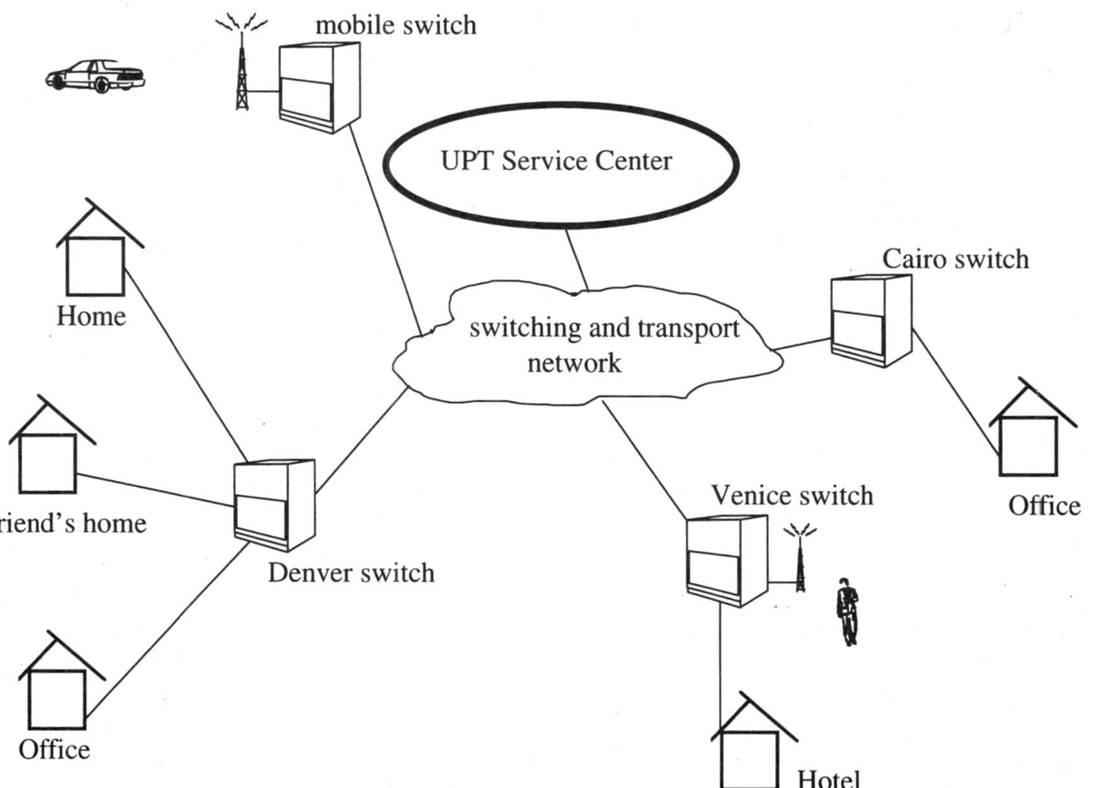

Figure 14.1 General UPT architecture.

of attention is being given to these kinds of future services by manufacturers, service providers, and investors. A service such as the one described below may be just around the corner.

Figure 14.1 shows the general structure of a UPT network. It consists of a UPT service center and a switching and transport network. The UPT service center is responsible for the storage of all information pertaining to a subscriber: the profile, current registration and current location if registered on the wireless subnetwork, security information, and anything else which is necessary to provide the services available to subscribers. The UPT service center is a network of different nodes performing functions that are similar to HLR, service control point (SCP), and service node (SN). It has its own structure which allows the information to be found, updated, and retrieved. It is a worldwide network and therefore covers subscribers everywhere.

The switching and transport network is used by the UPT service center to convey information, set up calls, and provide general connectivity among different subscribers and devices. It is a worldwide network designed with a hierarchical structure for an optimized use. It consists of switching and transmission equipment. We can speculate that convergence of computing and telecommunications would create new network elements which could perform both switching and computing functions. The advances of the Internet and other technological developments in this area are clearly pointing that way. However, there will still be functions oriented toward switching and functions oriented toward computing, so we can still use the same functional example to observe the behavior of the total network. Let's consider an example based on Fig. 14.1.

A UPT subscriber (Mr. Smith) has a personal number (500-123-4567). He has one wireless handset and a variety of other communications devices in different places. These devices are fax machines and computers. Mr. Smith carries his handset with him most of the time and has a personal computer at home, a fax machine and a personal computer in his Denver office, and a fax machine in his Cairo office. In our examples we will also see Mr. Smith in his friend's house and, while on vacation, in a Venice hotel.

When a UPT subscriber registers, a service profile is created in the network. For Mr. Smith the profile is as follows:

- The name and billing address
- A list of physical devices, their locations, and capabilities
- Mobile handset characteristics
- A list of voice mailboxes:

   - ▲ Denver office
   - ▲ Home
   - ▲ Cairo office

- ■ Billing currency: $$
- ■ Call delivery defaults:
   - ▲ All phone calls to be delivered to a wireless handset, unless otherwise specified.
   - ▲ Fax calls originated in the Western Hemisphere are to be delivered to the fax machine in Denver.
   - ▲ Fax calls originated in the Eastern Hemisphere are to delivered to the fax machine in Cairo. Notification of this delivery to be sent to the Denver office.
   - ▲ Phone calls after 6 p.m., except from numbers X, Y, and Z, are to be delivered to the home PC as e-mail.
   - ▲ If a phone call from number W arrives after 6 p.m., the notification to the mobile handset is to be always issued.

In this example we will follow Mr. Smith and some of his communications. His day starts early and soon he is in his car on the road, at which time someone makes a phone call to number 500-123-4567. The sequence of events for this call delivery is as follows (see Fig. 14.2).

1. A call is originated.
2. A switching network requests the UPT service center for assistance in finding the destination for the call.
3. The service center informs the switching network that the subscriber is registered as a mobile user and is currently served by a particular mobile switch.
4. The call is routed to the mobile switch.
5. The switch delivers the call to Mr. Smith.

While this call is being set up, a fax from Tokyo is sent to the same number – 500-123-4567. Figure 14.3 shows the activities in this case.

1. A fax call is originated in Tokyo.
2. The UPT service center is informed.
3. According to Mr. Smith's profile, such a call is to be delivered to Cairo.

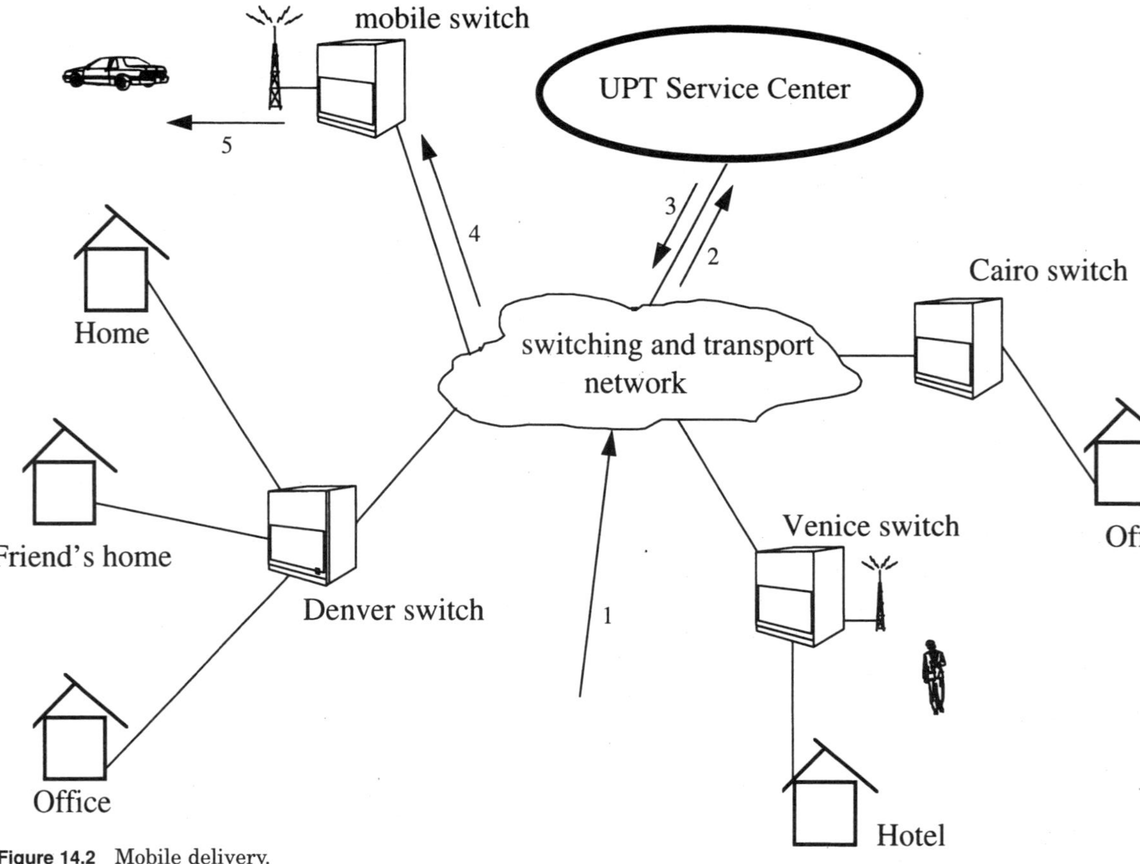

Figure 14.2 Mobile delivery.

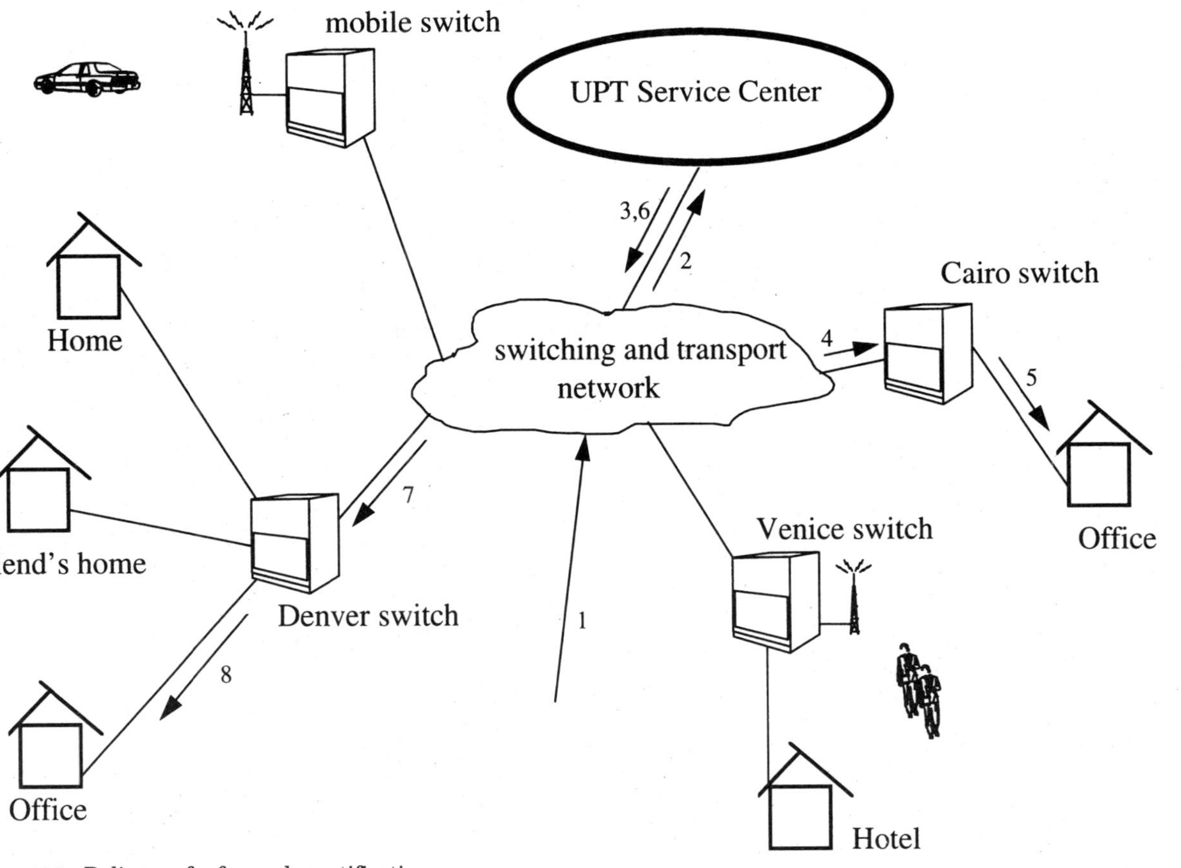

Figure 14.3 Delivery of a fax and a notification.

4. The switching network delivers the call to the Cairo switch.
5. The Cairo switch delivers the call to the fax machine.
6. The UPT service center also issues a notification about the fax to the Denver office as e-mail.
7. The notification is delivered to the Denver switch.
8. The notification is delivered to the office PC.

Mr. Smith arrives in his office in Denver. First he registers as a wireline user located in his office, then he checks his PC and finds the notification about the fax from Tokyo received at a particular time. In the meantime, a new call is originated to the same directory number. The activities related to this call are shown in Fig. 14.4.

1. A call is originated.
2. The UPT service center is informed.
3. The UPT service center directs the call to be delivered to Mr. Smith's office telephone device.
4. The call is delivered to the Denver switch.
5. The call is delivered to Mr. Smith's office. Mr. Smith is alerted, screens the call, and finds that it is a low-priority call from a relative.
6. Mr. Smith directs the network to redirect this call to his home mailbox.
7. The call is delivered according to the instructions.

Now Mr. Smith goes home for lunch. He registers at his home phone, makes phone calls, and does not receive any. He deregisters and moves to his friend's house. At this time another call to his personal number has been originated. Because no explicit wireline registration exists, the call is forwarded to his mobile phone. Assuming that the same mobile switch can serve Mr. Smith as previously, the activities shown in Fig. 14.2 are applicable to this latest case. Mr. Smith answers the call on his wireless handset in his friend's house. After visiting with a friend, Mr. Smith drives to the airport and flies to Venice. After his arrival another call is originated for him. Figure 14.5 shows the activities in this case.

1. A call is originated.
2. The UPT service center is informed.

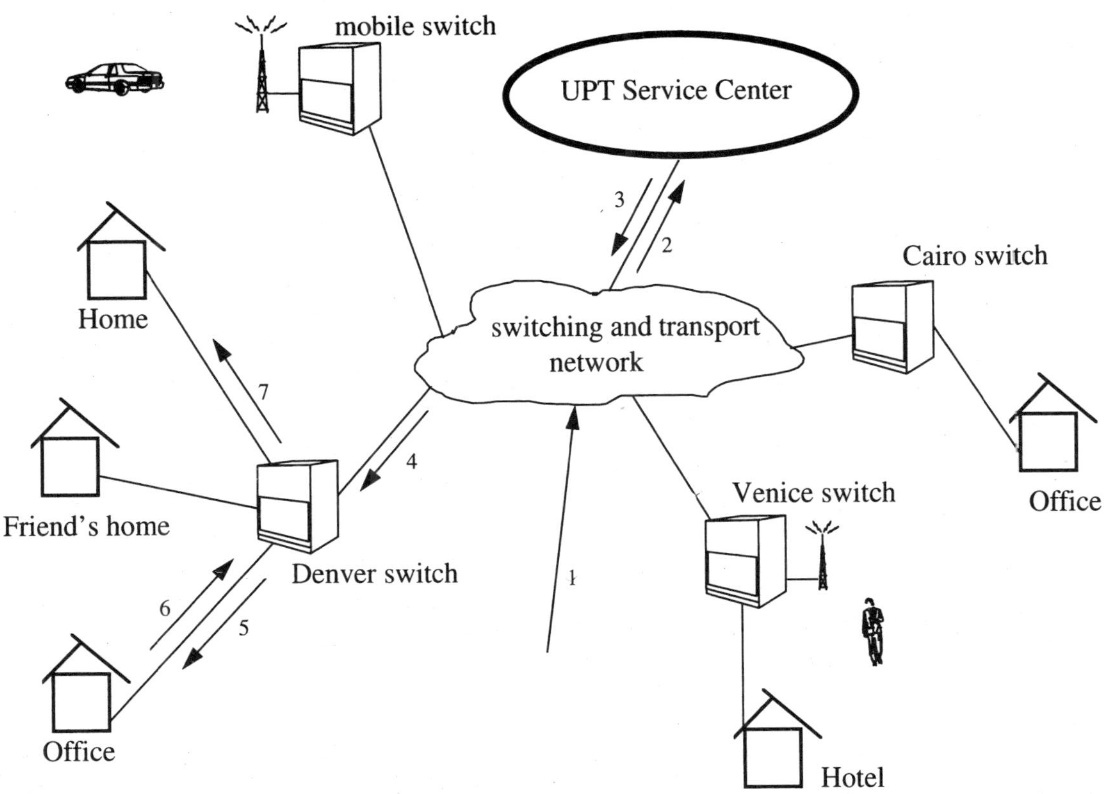

Figure 14.4 Redirection of a delivered call.

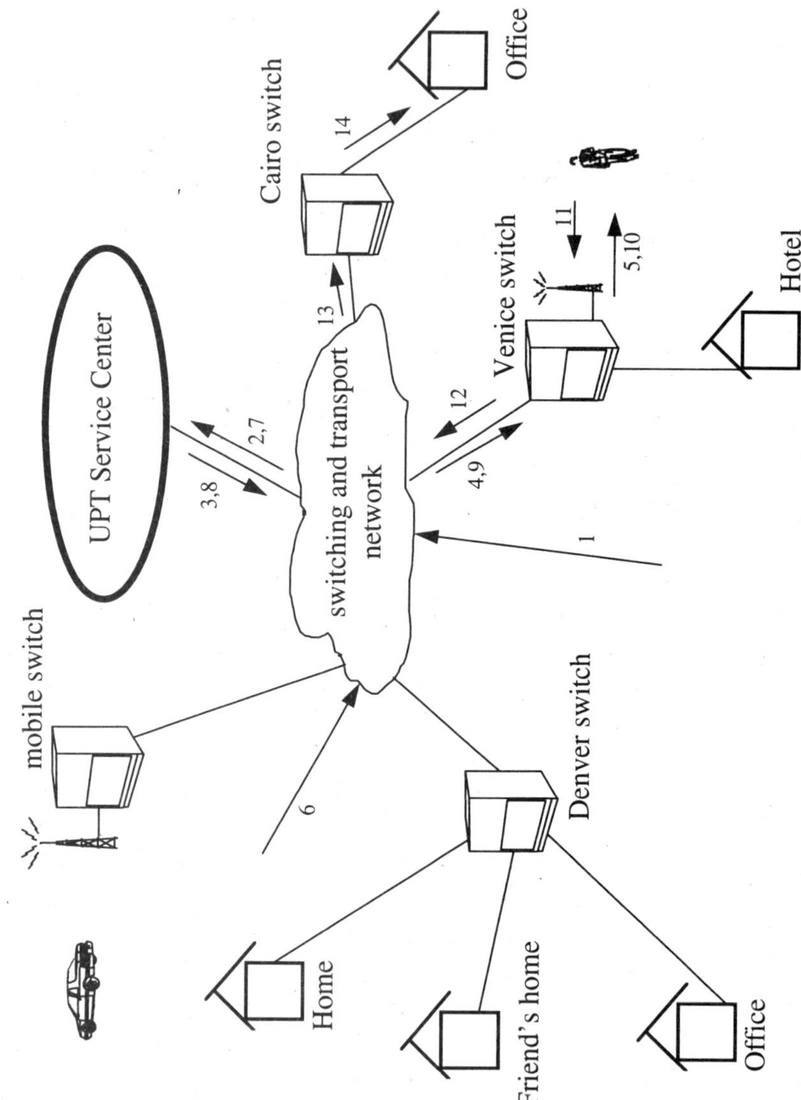

Figure 14.5 Call delivery away from home.

3. The UPT service center identifies that the call is to be delivered to the wireless handset and locates Mr. Smith in Venice.
4. The call is delivered to the Venice switch, which serves both wireline and wireless subscribers.
5. The call is delivered to Mr. Smith.
6. Another call is originated to Mr. Smith's personal number.
7. The UPT service center is informed.
8. The UPT service center directs the call to be delivered in Venice.
9. The call is delivered to the Venice switch.
10. The call is delivered to Mr. Smith. He is alerted to the second call (call waiting service) and screens this call.
11. Mr. Smith directs the second call to be forwarded to his Cairo office.
12. The call is returned to the switching network.
13. The call is delivered to the Cairo switch.
14. The call is delivered to the office.

The examples discussed here are presented only to demonstrate the kinds of capabilities that will be available to subscribers when UPT becomes a reality. It is difficult to project when this will occur, but it is certain that the capabilities will grow in stages. This process has already started, but a complete UPT environment requires developments in a number of different areas. Even our relatively simple examples show that different systems constituting the UPT service center would have to handle many messages, practically for every call. This requires powerful computers and an ability to store a lot of data. Without such computing support, UPT cannot happen. The switching network has to be able to transport a lot of information among a large number of different network nodes. In this area Asynchronous Transfer Mode (ATM) transmission and switching should help. Another necessary area is worldwide standards, whether formal or de facto. The standards are absolutely essential since they will allow straightforward connectivity between various networks in different parts of the world. Standards are necessary in the areas of signaling, application protocols, information transport, and services. The promise of UPT is to provide the same services in any part of the world, and without standards such a goal would be difficult to achieve.

Although a lot has to be done to reach the goal of UPT, it is achievable. As with many examples of other advances in life, it may happen sooner than we expect.

Abbreviations

AC	Authentication Center
ACSE	Association Control Service Element
AIN	Advanced Intelligent Network
AM	Access Manager
AMA	Automatic Message Accounting
AMPS	Advanced Mobile Phone System
ANSI	American National Standards Institute
ASE	Application Service Element
ASN.1	Abstract Syntax Notation One
AT	Access Tandem
ATM	Asynchronous Transfer Mode
BISDN	Broad-band ISDN
BML	Business Management Layer
BRI	Basic Rate Interface
BS	Base Station
BSC	Base Station Controller
BSS	Base Station Subsystem
BSSMAP	Base Station Subsystem Management Part
BTA	Basic Trading Area
BTS	Base Transceiver Station
CAVE	Cellular Authentication and Voice Encryption
CCS	Common Channel Signaling
CDMA	Code Division Multiple Access
CLASS	Custom Local Area Switching Services
CM	Communication Management
CMIP	Communication Management Information Protocol

CMIS	Communication Management Information Service
CMISE	Communication Management Information Service Element
CNAM	Calling Name Delivery
CND	Calling Number Delivery
CO	Central Office
CPE	Customer Premise Equipment
CTIA	Cellular Telecommunications Industry Association
DECT	Digital European Cordless Telecommunications
DF	Distribution Function
DMH	Data Message Handling
DN	Directory Number
DSS1	Digital Subscriber System 1
DTAP	Direct Transfer Application Part
DTMF	Dual-Tone Multifrequency
EDP	Event Detection Point
EIR	Equipment Identity Register
EML	Element Management Layer
EO	End Office
ESN	Electronic Serial Number
ETSI	European Telecommunications Standards Institute
FCC	Federal Communications Commission
GMSC	Gateway Mobile Switching Center
GR	Generic Requirement
GSM	Global System Mobile
GTT	Global Title Translation
HLR	Home Location Register
IAM	Initial Address Message
IMEI	International Mobile Equipment Identity
IMSI	International Mobile Subscriber Identification
IN	Intelligent Network
INAP	Intelligent Network Application Protocol
INPA	Interchangeable Numbering Plan Area
IP	Intelligent Peripheral
IS	Interim Standard
ISDN	Integrated Services Digital Network

ISM	Industrial, Scientific, and Medical
ISNI	Intermediate Signaling Network Identification
ISO	International Standards Organization
ISUP	ISDN Services User Part
ITU	International Telecommunication Union
IWF	Interworking Function
IXC	Interexchange Carrier
JTC	Joint Technical Committee
LAPD	Link Access Protocol on the D-Channel
LATA	Local Access and Transport Area
LEC	Local Exchange Carrier
LT	Local Tandem
MAP	Mobile Application Part
MC	Message Center
MIN	Mobile Identification Number
MM	Mobility Management
MMAP	Mobility Management Application Protocol
MS	Mobile Station
MSA	Metropolitan Statistical Area
MSC	Mobile Switching Center
MSISDN	Mobile Station ISDN Number
MTA	Major Trading Area
MTP	Message Transfer Part
MTSO	Mobile Telephone Switching Office
MVLR	Master Visitor Location Register
NANPA	North American Numbering Plan Administrator
NA/WCPE	North American Wireless Customer Premise Equipment
NCAS	Non-Call-Associated Signaling
NE	Network Element
NEL	Network Element Layer
NML	Network Management Layer
NPA	Numbering Plan Area
OAM+P	Operation, Administration, Maintenance, and Provisioning
OMAP	Operation, Maintenance, and Administration Part
OS	Operations System

OSI	Open System Interconnection
PACS	Personal Access Communications System
PBX	Private Branch Exchange
PCS	Personal Communications Services
PCSC	Personal Communications Switching Center
PHP	Personal Handy Phone
PHS	Personal Handy System
PIC	Point in Call
PLMN	Public Land Mobile Network
PMC	Personal Mobility Controller
PMD	Personal Mobility Data-Store
PSISDN	Personal Station ISDN Number
PSRN	Personal Station Roaming Number
POTS	Plain Old Telephone Service
PRI	Primary Rate Interface
PS	Personal Station
PSC	PCS Switching Center
PSTN	Public Switched Telephone Network
RACS	Radio Access System Controller
RBOC	Regional Bell Operating Company
RP	Radio Port
RPC	Radio Port Controller
RPI	Radio Port Intermediary
RSA	Rural Statistical Area
ROSE	Remote Operations Service Element
RR	Radio Resource
RS	Radio System
SCCP	Signaling Connection Control Part
SCP	Service Control Point
SCE	Service Creation Environment
SIB	Service-Independent Building Blocks
SIM	Subscriber Identity Module
SME	Short Message Entity
SML	Service Management Layer
SMS	Service Management System

SMS	Short Message Service
SN	Service Node
SRD	Service Requirements Document
SSD	Shared Secret Data
SS7	Signaling System #7
SSP	Service Switching Point
STP	Signal Transfer Point
SU	Subscriber Unit
TAG	Technology Ad-Hoc Group
TC	Toll Center
TCAP	Transaction Capabilities Application Part
TDMA	Time Division Multiple Access
TDP	Trigger Detection Point
TE	Terminal Equipment
TIA	Telecommunications Industry Association
TLDN	Temporary Line Directory Number
TMC	Terminal Mobility Controller
TMD	Terminal Mobility Data-Store
TMN	Telecommunications Management Network
TMSI	Temporary Mobile Subscriber Identity
TR	Technical Requirement
TSC	Technical Subcommittee
TUP	Telephone User Part
VLR	Visitor Location Register
UPT	Universal Personal Telecommunications
VSAT	Very Small Aperture Terminal
WACS	Wireless Access Communication System
WIN	Wireless Intelligent Network
WVLR	Wireline Visitor Location Register

Bibliography

ANSI T1.651; Mobility Management Applications Protocol (MMAP).

ANSI T1.704; Stage 2 Service Description for Personal Communications Services—Circuit-Mode Switched Bearer Services.

Calhoun, George, *Wireless Access and the Local Telephone Network,* Artech House, Boston, 1992.

GR-246-CORE; *Bell Communications Research Specifications of Signaling System Number 7.*

GR-1129-CORE; *Advanced Intelligent Network (AIN) 0.2 Switch-Intelligent Peripheral Interface (IPI) Generic Requirements.*

GR-1280-CORE; *Advanced Intelligent Network (AIN) Service Control Point (SCP) Generic Requirements.*

GR-1286-CORE; *Advanced Intelligent Network (AIN) Operations Systems (OS)-Service Control Point (SCP) Interface Generic Requirements.*

GR-1298-CORE; *Advanced Intelligent Network (AIN) Switching Systems Generic Requirements.*

GR-1299-CORE; *Advanced Intelligent Network (AIN) Switch-Service Control Point (SCP)/Adjunct Interface Generic Requirements.*

INSIGHT Research Corporation, *Personal Wireless Communications,* 1991.

ITU-T Recommendation E.164; Numbering Plan for the ISDN Era.

ITU-T Recommendation E.212; Identification Plan for Land Mobile Stations.

ITU-T Recommendation E.213; Telephone and ISDN Numbering Plan for Land Mobile Stations in Public Land Mobile Networks (PLMN).

ITU-T E.215; Telephone/ISDN Numbering Plan for the Mobile-Satellite Services of INMARSAT.

ITU-T Recommendation Q.701; Functional Description of the Message Transfer Part (MTP) of Signaling System No. 7.

ITU-T Recommendation Q.705; Signaling Network Structure.

ITU-T Recommendation Q.711; Functional Description of Signaling Connection Control Part.

ITU-T Recommendation Q.761; Signaling System # 7—ISDN User Part.

ITU-T Recommendation Q.771; Functional Description of Transaction Capabilities.

ITU-T Recommendation Q.772; Transaction Capabilities Information Element Definitions.

ITU-T Recommendation Q.773; Transaction Capabilities Formats and Encoding.

ITU-T Recommendation Q.774; Transaction Capabilities Procedures.

ITU-T Recommendation Q.775; Guidelines for Using Transaction Capabilities.

ITU-T Recommendation Q.811; Specification of TMN Interfaces, Q3 and X Interfaces, Lower Layer Protocol Profiles.

ITU-T Recommendation Q.812; Specification of TMN Interface Protocols, Q3 and X Interfaces, Upper Layer Protocol Profiles.

ITU-T Recommendation Q.920-Q.921; DSS1 Data Link Layer.

ITU-T Recommendation Q.922; ISDN Data Link Layer Specification for Frame Mode Bearer.

ITU-T Recommendation Q.931; ISDN User-Network Interface Layer 3 Specification for Basic Call Control.

ITU-T Recommendation Q.932; Generic Procedures for the Control of ISDN Supplementary Services.

ITU-T Recommendation Q.1001; General Aspects of Public Land Mobile Networks.

ITU-T Recommendation Q.1200; Q-Series Intelligent Network Recommendation Structure.

ITU-T Recommendation Q.1201; Principles of Intelligent Network Architecture.

ITU-T Recommendation Q.1202; Intelligent Network Service Plane Architecture.

ITU-T Recommendation Q.1203; Intelligent Network Global Functional Plane Architecture.

ITU-T Recommendation Q.1204; Intelligent Network Distributed Functional Plane.

ITU-T Recommendation Q.1205; Intelligent Network Physical Plane Architecture.

ITU-T Recommendation Q.1208; General Aspects of the Intelligent Network Application Protocol.

ITU-T Recommendation Q.1211; Introduction to Intelligent Network Capability Set 1.

ITU-T Recommendation Q.1213; Global Functional Plane for Intelligent Network CS-1.

ITU-T Recommendation Q.1214; Distributed Functional Plane for Intelligent Network CS-1.

ITU-T Recommendation Q.1215; Physical Plane for Intelligent Network CS-1.

ITU-T Recommendation Q.1218; Interface Recommendations for Intelligent Network CS-1.

ITU-T Recommendation Q.1290; Glossary of Terms and Definitions of Intelligent Networks.

ITU-T Recommendation X.25; Interface Between Data Terminal Equipment (DTE) and Data Circuit-Terminating Equipment (DCE) for Terminals Operating in the Packet Mode and Connected to the Public Data Networks for Dedicated Circuits.

ITU-T Recommendation X.200; Reference Model of Open Systems Interconnection.

ITU-T Recommendation X.208; Specification of Abstract Syntax Notation One (ASN.1).

ITU-T Recommendation X.209; Specification of Basic Encoding Rules for ASN.1.

ITU-T Recommendation X.217; Service Definition for the Association Control Service Element.

ITU-T Recommendation X.219; Remote Operations: Model, Notation, and Service Definition.

ITU-T Recommendation X.227; Association Control Protocol Specification for Open Systems Interconnection.

ITU-T Recommendation X.229; Remote Operations: Protocol Specification.

Keiser, Bernard E., and Eugene Strange, *Digital Telephony and Network Integration,* Van Nostrand Reinhold, New York, 1985.

Mouly, Michel, and Marie-Bernadette Pautet, *The GSM System for Mobile Communications*, 1992.

Redl, Siegmund, et al., *An Introduction to GSM,* Artech House,Boston, 1995.

SR-TSV-002275, issue 2, 1994; *BOC Notes on the LEC Networks —1994.*

Stallings, William, *SNMP, SNMPv2, and CMIP: The Practical Guide to Network-Management Standards,* Addison-Wesley, Reading, MA, 1993.

Talley, David, *Basic Electronic Switching for Telephone Systems,* Hayden Book Company, Rochelle Park, N. J., 1982.

TR-NWT-001127; *Advanced Intelligent Network (AIN) Adjunct Generic Requirements.*

TR-NWT-001254; *Advanced Intelligent Network (AIN) OS/Adjunct Interface Generic Requirements.*

TR-NWT-001284; *Advanced Intelligent Network (AIN) 0.1 Switching Systems Generic Requirements.*

TR-NWT-001285; *Advanced Intelligent Network (AIN) 0.1 Switch-Service Control Point (SCP) Application Protocol Interface Generic Requirements;*

TIA IS-41; Cellular Radiotelecommunications Intersystem Operations.

TIA IS-652; PCN to PCN Intersystem Operations Based on DCS-1900.

TIA IS-653; ISDN-based A-Interface (radio system—PCSC) for 1800-MHz Personal Communications Systems.

Index